Springer Theses

Recognizing Outstanding Ph.D. Research

For further volumes:
http://www.springer.com/series/8790

Aims and Scope

The series "Springer Theses" brings together a selection of the very best Ph.D. theses from around the world and across the physical sciences. Nominated and endorsed by two recognized specialists, each published volume has been selected for its scientific excellence and the high impact of its contents for the pertinent field of research. For greater accessibility to non-specialists, the published versions include an extended introduction, as well as a foreword by the student's supervisor explaining the special relevance of the work for the field. As a whole, the series will provide a valuable resource both for newcomers to the research fields described, and for other scientists seeking detailed background information on special questions. Finally, it provides an accredited documentation of the valuable contributions made by today's younger generation of scientists.

Theses are accepted into the series by invited nomination only and must fulfill all of the following criteria

- They must be written in good English.
- The topic should fall within the confines of Chemistry, Physics, Earth Sciences, Engineering and related interdisciplinary fields such as Materials, Nanoscience, Chemical Engineering, Complex Systems and Biophysics.
- The work reported in the thesis must represent a significant scientific advance.
- If the thesis includes previously published material, permission to reproduce this must be gained from the respective copyright holder.
- They must have been examined and passed during the 12 months prior to nomination.
- Each thesis should include a foreword by the supervisor outlining the significance of its content.
- The theses should have a clearly defined structure including an introduction accessible to scientists not expert in that particular field.

Manas A. Pathak

Privacy-Preserving Machine Learning for Speech Processing

Doctoral Thesis accepted by
Carnegie Mellon University, USA

 Springer

Author
Dr. Manas A. Pathak
Carnegie Mellon University
Pittsburgh, PA
USA

Supervisor
Prof. Dr. Bhiksha Raj
Carnegie Mellon University
Pittsburgh, PA
USA

ISSN 2190-5053   ISSN 2190-5061   (electronic)
ISBN 978-1-4899-9120-1   ISBN 978-1-4614-4639-2 (eBook)
DOI 10.1007/978-1-4614-4639-2
Springer New York Heidelberg Dordrecht London

Printed on acid-free paper

Springer is part of Springer Science+Business Media (www.springer.com)

*Dedicated to my parents, Dr. Ashok and
Dr. Sulochana Pathak*

Supervisor's Foreword

Speech has been the primary mode of communication between humans since time immemorial. It has also historically been considered one of the most private forms of communication: it is considered socially, and often legally unacceptable to eavesdrop on private conversations.

And yet, when it comes to communicating with computers, people generally pay little heed to the potential loss of privacy. They freely utilize a variety of speech-based services, including information engines such as SIRI and voice authentication services, and cheerfully upload recordings of themselves into repositories such as YouTube, without considering the security implications.

In a recent article in MIT's Tech Review dated 28 June 2012, reporter David Talbot points out that a person's speech recordings carry biometric signatures. By casually using SIRI, for instance, users are providing the service with these signatures which could be used to gather information about the users themselves. Although not explicitly listed out in the article, these could include factors such as their identity, gender, age, nationality, emotional state, or even factors relating to health, none of which the user has consciously chosen to reveal to the system. More direct risks include editing the speech recordings to create fake recordings, or tracking the user's online habits by identifying them in online audio and multimedia repositories such as YouTube.

Privacy issues also crop up in other situations. Doctors, lawyers, or government agencies who would like to use publicly available speech recognition services, for instance, cannot do so due to legal restrictions on sharing the speech recordings. Data in speech data warehouses may similarly not be mined for privacy concerns.

On the other hand, the literature on the topic of how computer processing of speech may be performed without exposing users to the risks of privacy violation or unintended information leakage remains woefully small. The problem has largely been treated as one that might be addressed through policy. Alternately, it has been viewed, rather na, as one that might be handled by morphing speech through signal processing techniques or through lossy feature extraction from speech signals. Needless to say, these are weak or even incorrect solutions.

In this first-of-its-kind dissertation, Dr. Manas A. Pathak addresses the issue of privacy and security in speech-processing systems as a computational challenge. Encryption techniques have progressed to the point where it is possible to manipulate encrypted data in many ways without decrypting them. Other methods have been developed to convert variable-length stochastic data such as speech recordings into password-like strings. Through the use of these and other similar techniques, Dr. Pathak demonstrates how traditional speech-processing solutions could be made "secure"—affording inviolable guarantees to the user's privacy.

As the use of automated speech-processing technologies increases, the problems that this dissertation addresses may be expected to become increasingly prominent. As one of the first to identify the problem and propose solutions, we expect this dissertation to be highly influential in years to come. We may expect the ideas and techniques presented here to not only influence future researchers in speech privacy, but also to provide a foundation that can be built upon.

We expect this document to be useful to both established researchers and entry-level graduate students looking for problems to solve. To the established researchers in speech processing, this document will present the privacy challenges that current technologies present and give them an idea of approaches that may be taken to address them. To the beginning graduate student, this will provide a starting point, upon which additional research may be built.

Pittsburgh, August 2012   Dr. Bhiksha Raj

Acknowledgments

I would like to thank my advisor, Bhiksha Raj, with whom I have had the pleasure of working with during the course of my Ph.D. program at Carnegie Mellon University. Bhiksha is truly a visionary, and has made many contributions to multiple research areas: speech and signal processing, machine learning, privacy, optimization, dialog systems, knowledge representation, to name a few—he is symbolic of the interdisciplinary culture that LTI and CMU are known for. I learned a lot from working with Bhiksha: his depth of knowledge, his effective communication of ideas, and most importantly, his zeal to push the boundaries and advance the state of art. I appreciate the latitude he gave me in pursuing research, while providing invaluable guidance and support.

I also learned a lot from Rita Singh as well, who is also a visionary and an extraordinary researcher. Apart from the numerous useful technical discussions which I am thankful for, Rita and Bhiksha made myself and the other graduate students feel at home in Pittsburgh despite being literally thousands of miles away from our original homes. I thank my "grand-advisor" Rich Stern for his feedback that significantly helped me improve my research. I also thank Eric Nyberg who was my advisor during the masters program at CMU. It was with him that I obtained my first experience of doing graduate-level research and he always took special efforts in improving my speaking and writing skills.

I was fortunate in having an amazing thesis committee with Alan Black, Anupam Datta, Paris Smaragdis, and Shantanu Rane. I thank Alan for the useful guidance and suggestions at each stage of my research. I also learned a lot about privacy and security from being the teaching assistant for the seminar course instructed by him and Bhiksha. I thank Anupam for helping me in making the thesis much more rigorous through insightful discussions. His feedback during the proposal stage helped me vastly in focusing on relevant aspects and improving the quality of the thesis. Paris, of course, inspired me with the idea for this thesis through his multiple papers on the subject. I also thank him for his valuable guidance which especially helped me in seeing the big picture. I am grateful to Shantanu for mentoring me during my internship at MERL. It was a very fulfilling experience and was indeed the point at which my research got accelerated.

I also thank Shantanu for his suggestions which helped me make the thesis more comprehensive. I am also thankful to my other internship mentors: Wei Sun and Petros Boufounos at MERL, Richard Chow and Elaine Shi at PARC.

Special thanks to my former and current office-mates: Mehrbod Sharifi, Diwakar Punjani, Narjes Sharif Razavian during my masters program, Amr Ahmed during the middle years, Michael Garbus and Yubin Kim during the last year, and my co-members of the MLSP research group: Ben Lambert, Sourish Chaudhuri, Sohail Bahmani, Antonio Juarez, Mahaveer Jain, Jose Portelo, John McDonough, and Kenichi Kumatani, and other CMU graduate students. I enjoyed working with Mehrbod; it was my discussions with him that led to some of the important ideas in this thesis. Apart from being close friends, Diwakar originally motivated me towards machine learning and Amr taught me a lot about research methodologies, which I thank them for. I thank Sohail for being my first collaborator in my research on privacy. Thanks to Sourish, Ben, and Michael for the thought-provoking discussions.

Finally, I am thankful to my father Dr. Ashok Pathak and my mother Dr. Sulochana Pathak for their endless love and support. I dedicate this thesis to them.

Pittsburgh, July 2012   Manas A. Pathak

Contents

Acronyms

EER	Equal Error Rate
EM	Expectation Maximization
GMM	Gaussian Mixture Model
HE	Homomorphic Encryption
HMM	Hidden Markov Model
LSH	Locality Sensitive Hashing
MAP	Maximum *a* posteriori
MFCC	Mel Frequency Cepstral Coefficients
OT	Oblivious Transfer
PPSI	Privacy-preserving Speaker Identification
PPSV	Privacy-preserving Speaker Verification
SMC	Secure Multiparty Computation
TTP	Trusted Third Party
UBM	Universal Background Model
ZKP	Zero-Knowledge Proof

Part I
Introduction

Chapter 1
Thesis Overview

1.1 Motivation

Speech is one of the most private forms of personal communication. A sample of a person's speech contains information about the gender, accent, ethnicity, and the emotional state of the speaker apart from the message content. Speech processing technology is widely used in biometric authentication in the form of speaker verification. In a conventional speaker verification system, the speaker patterns are stored without any obfuscation and the system matches the speech input obtained during authentication with these patterns. If the speaker verification system is compromised, an adversary can use these patterns to later impersonate the user. Similarly, speaker identification is also used in surveillance applications. Most individuals would consider unauthorized recording of their speech, through eavesdropping or wiretaps as a major privacy violation. Yet, current speaker verification and speaker identification algorithms are not designed to preserve speaker privacy and require complete access to the speech data.

In many situations, speech processing applications such as speech recognition are deployed in a client-server model, where the client has the speech input and a server has the speech models. Due to the concerns for privacy and confidentiality of their speech data, many users are unwilling to use such external services. Even though the service provider has a privacy policy, the client speech data is usually stored in an external repository that may be susceptible to being compromised. The external service provider is also liable to disclose the data in case of a subpoena. It is, therefore, very useful to have privacy-preserving speech processing algorithms that can be used without violating these constraints.

M. A. Pathak, *Privacy-Preserving Machine Learning for Speech Processing*,
Springer Theses, DOI: 10.1007/978-1-4614-4639-2_1,

1.2 Thesis Statement

With the above motivation, we introduce privacy-preserving speech processing: algorithms that allow us to process speech data without being able to observe it. We envision a client-server setting, where the client has the speech input data and the server has speech models. Our privacy constraints require that the server should not observe the speech input and the client should not observe the speech models. To prevent the client and the server from observing each others input we require them to obfuscate their inputs. We use techniques from cryptography such as homomorphic encryption and hash functions that allow us to process obfuscated data through interactive protocols.

The main goal of this thesis is to show that:

"Privacy-preserving speech processing is feasible and useful."

To support this statement, we develop privacy-preserving algorithms for speech processing applications such as speaker verification, speaker identification, and speech recognition. We consider two aspects of feasibility: accuracy and speed. We create prototype implementations of each algorithm and measure the accuracy of our algorithms on standardized datasets. We also measure the speed by the execution time of our privacy-preserving algorithms over sample inputs and compare it with the original algorithm that does not preserve privacy. We also consider multiple algorithms for each task to enable us to obtain a trade-off between accuracy and speed. To establish utility, we formulate scenarios where the privacy-preserving speech processing applications can be used and outline the various privacy issues and adversarial behaviors that are involved. We briefly overview the three speech processing applications below.

A speaker verification system uses the speech input to authenticate the user, i.e., to verify if the user is indeed who he/she claims to be. In this task, a person's speech is used as a biometric. We develop a framework for privacy-preserving speaker verification, where the system is able to perform authentication without observing the speech input provided by the user and the user does not observe the speech models used by the system. These privacy criteria are important in order to prevent an adversary having unauthorized access to the user's client device or the system data from impersonating the user in another system. We develop two privacy-preserving algorithms for speaker verification. Firstly, we use Gaussian mixture models (GMMs) and create a homomorphic encryption-based protocol to evaluate GMMs over private data. Secondly, we apply locality sensitive hashing (LSH) and one-way cryptographic functions to reduce the speaker verification problem to private string comparison. There is a trade off between the two algorithms, the GMM-based approach provides high accuracy but is relatively slower, the LSH-based approach provides relatively lower accuracy, but is comparatively faster.

Speaker identification is a related problem of identifying the speaker from a given set of speakers that best corresponding to a given speech sample. This task finds use in surveillance applications, where a security agency such as the police has access to speaker models for individuals, e.g., a set of criminals it is interested in monitoring and an independent party such as a phone company might have access to the phone conversations. The agency is interested in identifying the speaker participating in a given phone conversation among its set of speakers, with a none of the above option. The agency can demand the complete recording from the phone company if it has a warrant for that person. By using a privacy-preserving speaker identification system, the phone company can provide the privacy guarantee to its subscribers that the agency will not be able to obtain any phone conversation for the speakers that are not under surveillance. Similarly, the agency does not need to send the list of speakers under surveillance to the phone company, as this itself is highly sensitive information and its leakage might interfere in the surveillance process. Speaker identification can be considered to be an extension of speaker verification to the multiclass setting. We extend the GMM-based and LSH-based approaches to create analogous privacy-preserving speaker identification frameworks.

Speech recognition is the task of converting a given speech sample to text. We consider a client-server scenario for speech recognition, where the client has a light-weight computation device to record the speech input and the server has the necessary speech models. In many applications, the client might not be comfortable in sharing its speech input with the server due to privacy constraints such as confidentiality of the information. Similarly, the server may not want to release its speech models as they too might be proprietary information. We create an algorithm for privacy-preserving speech recognition that allows us perform speech recognition while satisfying these privacy constraints. We create an isolated-word recognition system using hidden Markov models (HMMs), and use homomorphic encryption to create a protocol that allows the server to evaluate HMMs over encrypted data submitted by the client.

1.3 Summary of Contributions

To the best of our knowledge and belief, this thesis is the first end-to-end study of privacy-preserving speech processing. The technical contributions of this thesis along with relevant publications are as follows.

1. Privacy models for privacy-preserving speaker verification, speaker identification, and speech recognition.
2. GMM framework for privacy-preserving speaker verification (Pathak and Raj 2011, 2012a).
3. LSH framework for privacy-preserving speaker verification (Pathak and Raj 2012b).
4. GMM framework for privacy-preserving speaker identification (Pathak and Raj 2012a).
5. LSH framework for privacy-preserving speaker identification.
6. HMM framework for privacy-preserving isolated-keyword recognition (Pathak et al. 2011).

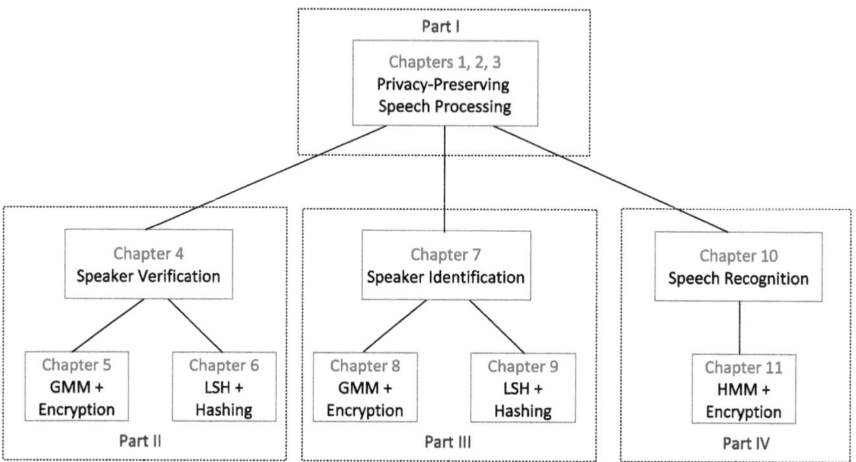

Fig. 1.1 Thesis organization

1.4 Thesis Organization

We summarize the thesis organization in Fig. 1.1. In Part I we overview the preliminary concepts of speech processing (Chap. 2) and privacy-preserving methodologies
(Chap. 3). In Part II, we introduce the problem of privacy-preserving speaker verification, and discuss the privacy issues in Chap. 4. We then present two algorithms
for privacy-preserving speaker verification: using GMMs in Chap. 5 and LSH in
Chap. 6. In Part III, we introduce the problem of privacy-preserving speaker identification with a discussion of the privacy issues in Chap. 7. We then present two
algorithms for privacy-preserving speaker identification: using GMMs in Chap. 8
and LSH in Chap. 9. In Part IV, we introduce the problem of privacy-preserving
speech recognition with a discussion of the privacy issues in Chap. 10 along with an
HMM-based framework for isolated-keyword recognition in Chap. 11. In Part V, we
complete the thesis by summarizing our conclusions in Chap. 12 and outlining future
work in Chap. 13.

References

Pathak MA, Raj B (2011) Privacy preserving speaker verification using adapted GMMs. Interspeech,
     In, pp 2405–2408
Pathak MA, Raj B (2012a) Privacy preserving speaker verification and identification using gaussian
     mixture models. In: IEEE transactions on audio, speech and, language processing (to appear)
Pathak MA, Raj B (2012b) Privacy preserving speaker verification as password matching. In: IEEE
     international conference on acoustics, speech and signal processing
Pathak MA, Rane S, Sun W, Raj B (2011) Privacy preserving probabilistic inference with hidden
     Markov models. In: IEEE international conference on acoustics, speech and signal processing

Chapter 2
Speech Processing Background

In this chapter, we review some of the building blocks of speech processing systems. We then discuss the specifics of speaker verification, speaker identification, and speech recognition. We will reuse these constructions when designing privacy-preserving algorithms for these tasks in the reminder of the thesis.

Almost all speech processing techniques follow a two-step process of signal parameterization followed by classification. This is shown in Fig. 2.1.

2.1 Tools and Techniques

2.1.1 Signal Parameterization

Signal parameterization is a key step in any speech processing task. As the audio sample in the original form is not suitable for statistical modeling, we represent it using *features*.

The most commonly used parametrization for speech is Mel-frequency cepstral coefficients (MFCC) (Davis and Mermelstein 1980). In this representation, we segment the speech sample into 25 ms windows, and take the Fourier transform of each window. This is followed by de-correlating the spectrum using a cosine transform, then taking the most significant coefficients.

If x is a frame vector of the speech samples, F is the Fourier transform in matrix form, M is the set of Mel filters represented as a matrix, and D is a DCT matrix, MFCC feature vectors can be computed as

$$MFCC(x) = D \log(M((Fx) \cdot \mathrm{conjugate}(Fx))).$$

M. A. Pathak, *Privacy-Preserving Machine Learning for Speech Processing*,
Springer Theses, DOI: 10.1007/978-1-4614-4639-2_2,
© Springer Science+Business Media New York 2013

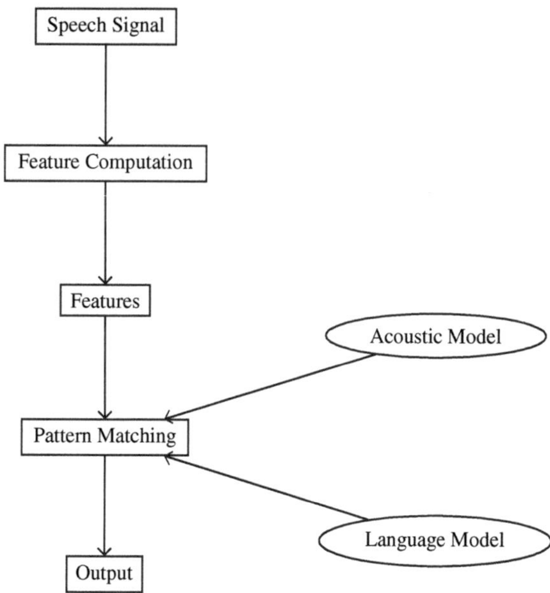

Fig. 2.1 Work flow of a speech processing system

2.1.2 Gaussian Mixture Models

Gaussian mixture models (GMMs) are commonly used generative models for density estimation in speech and language processing. The probability of the model generating an example is given by a mixture of Gaussian distributions (Fig. 2.2).

A GMM λ comprises of M multivariate Gaussians each with a mean and covariance matrix. If the mean vector and covariance matrix of the jth Gaussian are respectively μ_j and Σ_j, for an observation x, we have

$$P(x|\lambda) = \sum_j w_j \mathcal{N}(\mu_j, \Sigma_j),$$

where w_j are the mixture coefficients that sum to one. The above mentioned parameters can be computed using the expectation-maximization (EM) algorithm.

2.1.3 Hidden Markov Models

A hidden Markov model (HMM) (Fig. 2.3), can be thought of as an example of a Markov model in which the state is not directly visible but the output of each state can be observed. The outputs are also referred to as observations. Since observations

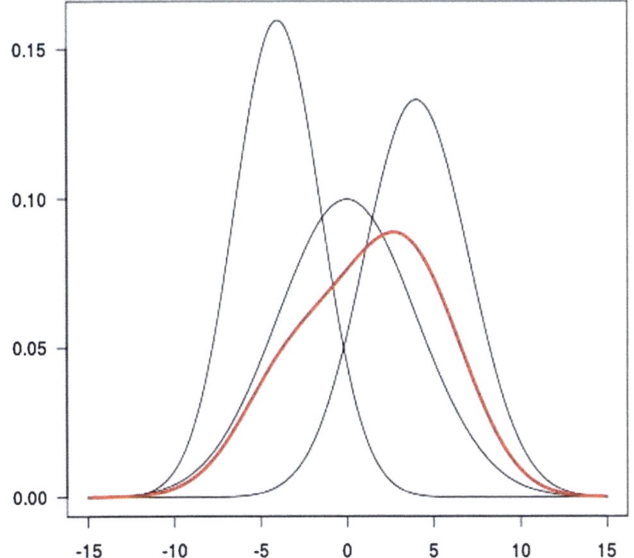

Fig. 2.2 An example of a GMM with three Gaussian components

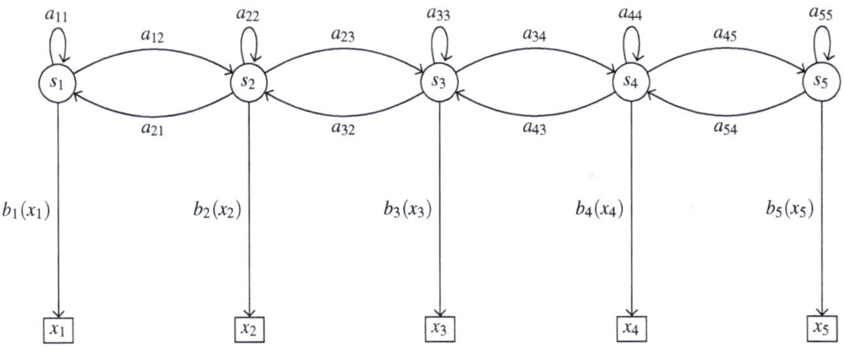

Fig. 2.3 An example of a 5-state HMM

depend on the hidden state, an observation reveals information about the underlying state.

Each HMM is defined as a triple $M = (A, B, \Pi)$, in which

- $A = (a_{ij})$ is the state transition matrix. Thus, $a_{ij} = \Pr\{q_{t+1} = S_j | q_t = S_i\}$, $1 \le i, j \le N$, where $\{S_1, S_2, \ldots, S_N\}$ is the set of states and q_t is the state at time t.
- $B = (b_j(v_k))$ is the matrix containing the probabilities of the observations. Thus, $b_j(v_k) = \Pr\{\mathbf{x}_t = v_k | q_t = S_j\}$, $1 \le j \le N$, $1 \le k \le M$, where $v_k \in V$ which is the set of observation symbols, and $\mathbf{x}_t$ is the observation at time t.

- $\Pi = (\pi_1, \pi_2, \ldots, \pi_N)$ is the initial state probability vector, that is, $\pi_i = \Pr\{q_1 = S_i\}$, $i = 1, 2, \ldots, N$.

Depending on the set of observation symbols, we can classify HMMs into those with discrete outputs and those with continuous outputs. In speech processing applications, we consider HMMs with continuous outputs where each of the observation probabilities of each state is modeled using a GMM. Such a model is typically used to model the sequential audio data frames representing the utterance of one sound unit, such as a phoneme or a word.

For a given sequence of observations $\mathbf{x}_1, \mathbf{x}_2, \ldots, \mathbf{x}_T$ and an HMM $\lambda = (A, B, \Pi)$, one problem of interest is to efficiently compute the probability $P(\mathbf{x}_1, \mathbf{x}_2, \ldots, \mathbf{x}_T | \lambda)$. A dynamic programming solution to this problem is the forward algorithm.

2.2 Speaker Identification and Verification

In speaker verification, a system tries to ascertain if a user is who he or she claims to be. Speaker verification systems can be text dependent, where the speaker utters a specific pass phrase and the system verifies it by comparing the utterance with the version recorded initially by the user. Alternatively, speaker verification can be text independent, where the speaker is allowed to say anything and the system only determines if the given voice sample is close to the speaker's voice. Speaker identification is a related problem in which we identify if a speech sample is spoken by any one of the speakers from our pre-defined set of speakers. The techniques employed in the two problems are very similar, enrollment data from each of the speakers are used to build statistical or discriminative models for the speaker which are employed to recognize the class of a new audio recording.

2.2.1 Modeling Speech

We discuss some of the modeling aspects of speaker verification and identification below. Both speaker identification and verification systems are composed of two distinct phases, a training phase and a test phase. The training phase consists of extracting parameters from the speech signal to obtain features and using them to train a statistical model. We typically use MFCC features of the audio data instead of the original samples as they are known to provide better accuracy for speech classification.

Given a speech sample y and a hypothesized speaker s, the speaker verification task can be formulated as determining if y was produced by S. In the speaker identification task, we have a pre-defined set of hypothesized speakers $S = \{S_0, S_1, \ldots, S_K\}$ and we are trying to identify which speaker $s \in S$ would have spoken the speech sample y. Speaker S_0 represents the none of the above case where the speech sample does not match any of the other K speakers. In this discussion, we consider the single-speaker case where y is assumed to have spoken by only one speaker, please refer to Dunn et al. (2000) for work on detection from multi-speaker speech.

Speaker identification is simply finding the speaker s^* having the highest probability of generating the speech sample.

$$s^* = \operatorname*{argmax}_{s \in S} P(y|s).$$

Speaker verification on the other hand can be modeled as the following hypothesis test:

H_0: y is spoken by the speaker s

H_1: y is not spoken by the speaker s

We use a likelihood ratio test (LRT) to decide between the two hypothesis.

$$\frac{P(y|H_0)}{P(y|H_1)} \begin{cases} > \theta & \text{accept } H_0, \\ < \theta & \text{accept } H_1, \end{cases} \tag{2.1}$$

where θ is the decision threshold. The main problem in speaker identification and verification is effectively modeling the probabilities $P(y|s)$ and $P(y|H_0)$.

Modeling for a given speaker s and the hypothesis H_0 is well defined and can be done using training data for that speaker. On the other hand the alternative hypothesis H_1 is open-ended as it represents the entire space of possible speakers except s. There are exactly the same issues with modeling the "none of the above" speaker S_0 in speaker identification. The main approach for modeling the alternative hypothesis is by collecting speech samples from several speakers and training a single model called the universal background model (UBM) (Carey et al. 1991; Reynolds 1997). One benefit of this approach in speaker verification is that a single UBM can be used as the alternate hypothesis for all speakers. There has been work on selection and composition of the right kind of speakers used to train the UBM and also to use multiple background models tailored for a specific set of speakers (Matsui and Furui 1995; Rosenberg and Parthasarathy 1996; Heck and Weintraub 1997).

Selection of the right kind of probability model is a very important step in the implementation of any speech processing task and is closely dependent on the features used and other application details. For text-independent speaker identification and verification, GMMs have been used very successfully in the past. For text-dependent tasks, the additional temporal knowledge can be integrated by using a HMM, but the use of such complex models have not been shown to provide any significant advantage over using GMMs (Bimbot et al. 2004).

For a GMM Φ with M mixture components, the likelihood function density for a D-dimensional feature vector $\mathbf{x}$ is given by

$$P(\mathbf{x}|\Phi) = \sum_{i=1}^{M} w_i \, P_i(\mathbf{x}),$$

where w_i are the mixture components and $P_i(\mathbf{x})$ is the multivariate Gaussian density parameterized by mean μ_i and covariance Σ_i. Given a collection of vectors $\mathbf{x}$ from the training data, the GMM parameters are estimated using expectation-maximization (EM) algorithm.

For recognition, we assume the feature vectors of a speech sample to be independent. The model probabilities are scaled by the number of feature vectors to normalize for the length of the sample. The log-likelihood of a model Φ for a sequence of feature vectors $X = \{x_1, \ldots, x_T\}$ is given by

$$\log P(X|\Phi) = \frac{1}{T} \log \prod_{t=1}^{T} P(\mathbf{x}_t|\Phi) = \frac{1}{T} \sum_{t=1}^{T} \log P(\mathbf{x}_t|\Phi).$$

Basic speaker verification and identification systems use a GMM classifier model trained over the voice of the speaker. In case of speaker verification, we train a binary GMM classifier using the audio samples of the speaker as one class and a universal background model (UBM) as another class (Campbell 2002). The UBM is trained over the combined speech data of all other users. Due to the sensitive nature of their use in authentication systems, speaker verification classifiers need to be robust to false positives. In case of doubt about the authenticity of a user, the system should choose to reject. In case of speaker identification, we also use the UBM to categorize a speech sample as not being spoken by anyone from the set of speakers.

In practice, we need a lot of data from one speaker to train an accurate speaker classification model and such data is difficult to acquire. Towards this, Reynolds et al. (2000) proposed techniques for maximum *a posteriori* adaptation to derive speaker models from the UBM. These adaptation techniques have been extended by constructing *supervectors* consisting of concatenated means of the mixture components (Kenny and Dumouchel 2004). The supervector formulation has also been used with support vector machine (SVM) classification methods. Other approaches for representing speech samples with noise robustness include factor analysis (Dehak et al. 2011). These methods can be incorporated in our privacy-preserving speaker verification framework.

2.2.2 Model Adaptation

In the above discussion, we considered GMM as a representation of speaker models. In practice, however, we have limited quantity of speech data from individual speakers. It is empirically observed that GMMs obtained from adapting the UBM to speaker data from individual speakers significantly outperform the GMMs trained directly on the speaker data (Bimbot et al. 2004; Reynolds 1997). We present the algorithm for maximum *a posteriori* (MAP) adaptation of the UBM.

The MAP adaptation procedure consists of a sample estimate of the speaker model parameters such as the mixture weights and the means, followed by their interpolation with the UBM. Given set of enrollment speech data frames $x_1, \ldots, x_T$, we first compute the *a posteriori* probabilities of the individual Gaussians in the UBM $\lambda_U = \{w_i^U, \mu_i^U, \Sigma_i^U\}$. For the ith mixture component of the UBM,

$$P(i|x_t) = \frac{w_i^U \, \mathcal{N}(x_t; \mu_i^U, \Sigma_i^U)}{\sum_j w_j^U \, \mathcal{N}(x_t; \mu_j^U, \Sigma_j^U)}. \tag{2.2}$$

Similar to the maximization step of EM, the *a posteriori* probabilities are then used to compute new weights, mean, and second moment parameter estimates.

$$\begin{aligned}
w_i' &= \frac{1}{T} \sum_t P(i|x_t), \\
\mu_i' &= \frac{\sum_t P(i|x_t) x_t}{\sum_t P(i|x_t)}, \\
\Sigma_i' &= \frac{\sum_t P(i|x_t) x_t x_t^T}{\sum_t P(i|x_t)}.
\end{aligned} \tag{2.3}$$

Finally, the parameters of the adapted model $\lambda_s = \{\hat{w}_i^s, \hat{\mu}_i^s, \hat{\Sigma}_i^s\}$ are given by a convex combination of these new parameter estimates and the UBM parameters as follows.

$$\begin{aligned}
\hat{w}_i^s &= \alpha_i w_i' + (1 - \alpha_i) w_i^U, \\
\hat{\mu}_i^s &= \alpha_i \mu_i' + (1 - \alpha_i) \mu_i^U, \\
\hat{\Sigma}_i^s &= \alpha_i \Sigma_i' + (1 - \alpha_i) \left[\Sigma_i^U + \mu_i^U \mu_i^{UT} \right] - \hat{\mu}_i^s \hat{\mu}_i^{sT}.
\end{aligned} \tag{2.4}$$

The adaptation coefficients α_i control the amount of contribution of the enrollment data relative to the UBM.

2.2.3 Supervectors with LSH

Campbell et al. (2006) extend the adapted GMM algorithm by constructing a *supervector* (SV) for each speech sample. The supervector is obtained by performing maximum *a posteriori* (MAP) adaptation of the UBM over a single speech sample and concatenating the means of the adapted model. Given the adapted model $\lambda_s = \{\hat{w}_i^s, \hat{\mu}_i^s, \hat{\Sigma}_i^s\}$ with M-mixture components, the supervector sv is given by $(\hat{\mu}_1^s \parallel \hat{\mu}_2^s \parallel \cdots \parallel \hat{\mu}_M^s)$.

This supervector is then used as a feature vector instead of the original frame vectors of the speech sample. The verification is performed using a binary support vector machine (SVM) classifier for each user trained on supervectors obtained from enrollment utterances as instances labeled according to one class and impostor data as instances labeled according to the opposite class. As the classes are usually not

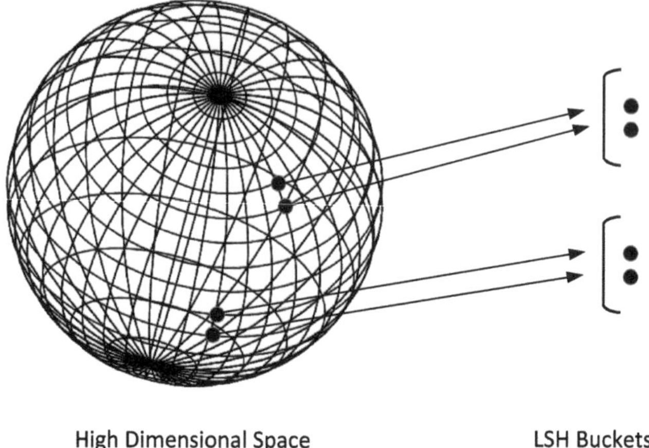

Fig. 2.4 LSH maps similar points to the same bucket

separable in the original space, (Campbell et al. 2006) also use a kernel mapping that is shown to achieve higher accuracy. Instead of using SVMs with kernels, we use k-nearest neighbors trained on supervectors as our classification algorithm. Our motivation for this are, firstly, k-nearest neighbors also perform classification with non-linear decision boundaries and are shown to achieve accuracy comparable to SVMs with kernels (Mariéthoz et al. 2009). Secondly, by using the LSH transformations we discuss below, we reduce the k-nearest neighbors computation to string comparison, which can be easily done with privacy without requiring an interactive protocol.

Locality Sensitive Hashing

Locality sensitive hashing (LSH) (Indyk and Motwani 1998) is a widely used technique for performing efficient approximate nearest-neighbor search. An LSH function $L(\cdot)$ proceeds by applying a random transformation to a data vector x by projecting it to a vector $L(x)$ in a lower dimensional space, which we refer to as the LSH *key* or *bucket*. A set of data points that map to the same key are considered as approximate nearest neighbors (Fig. 2.4).

As a single LSH function does not group the data points into fine-grained clusters, we use a hash key obtained by concatenating the output of k LSH functions. This k-bit LSH function $L(x) = L_1(x) \cdots L_k(x)$ maps a d-dimensional vector into a k-bit string. Additionally, we use m different LSH keys that are computed over the same input to achieve better recall. Two data vectors x and y are said to be neighbors if at least one of their keys, each of length k, matches exactly. One of the main advantages of using LSH is its efficiency: by precomputing the keys, the approximate nearest neighbor search can be done in time sub-linear to the number of instances in the dataset.

A family of LSH functions is defined for a particular distance metric. A hash function from this family has the property that data points, that are close to each other as defined by the distance metric, are mapped to the same key with high probability. There exist LSH constructions for a variety of distance metrics, including arbitrary kernels (Kulis and Grauman 2009), but we mainly consider LSH for Euclidean distance (E2LSH) (Datar et al. 2004) and cosine distance (Charikar 2002) as the LSH functions for these constructions are simply random vectors. As the LSH functions are data independent, it is possible to distribute them to multiple parties without the loss of privacy.

The LSH construction for Euclidean distance with k random vectors transforms a d-dimensional vector into a vector of k integers, each of which is a number between 0 and 255. The LSH construction for cosine distance with k random vectors similarly transforms the given data vector into a binary string of length k.

2.2.4 Reconstructing Data from LSH Keys

Due to the locality sensitive property, each LSH key provides information about the data point. We consider reconstructing the data point from LSH keys for cosine distance, but our analysis would hold for other distance metrics. A LSH function for cosine and Euclidean distances producing a k bit key is based on k random hyperplanes $\{r_1, \ldots, r_k\}$. Given a data point, we project it using each of the k random hyperplanes and determine the key by the side on which the data point lies.

$$
L_i(x) = \begin{cases} 1 & \text{if } r_i^T x \geq 0, \\ 0 & \text{if } r_i^T x < 0. \end{cases}
$$

A k bit LSH key provides k bits of entropy. Computing m LSH keys over the same data point reveals mk bits of information.

An LSH key is defined as the sector between two hyperplanes, this is because two vectors x and y lying in this space would have little angular distance $\theta(x, y)$, that corresponds to similarity in terms of cosine distance. By observing multiple keys computed using different LSH functions, we can further localize the data point by the space between any two hyperplanes, which may be from different LSH functions. We show an example of 2-dimensional LSH in Fig. 2.5, application of one LSH function localizes the data point in a wider space, but application of two LSH function further localizes the space between a hyperplane of the first and the second LSH function. As we mentioned in the analysis above, the degree of localization depends on the size of the LSH key, i.e., the number of hyperplanes, and the number of LSH keys. By observing a large number of LSH keys, we can localize a point to lie on a hyperplane.

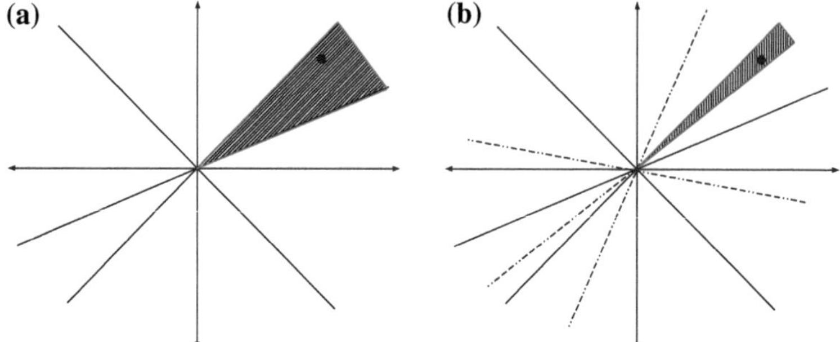

Fig. 2.5 Application of one and two LSH functions. **a** One LSH function (*solid lines*). **b** Two LSH functions (*solid* and *dashed lines*)

As we shall see in later chapters, the information revealed about the data point from the LSH keys causes a problem with respect to privacy. We satisfy the privacy constraints by obfuscating the LSH key. A simple way of doing that is by applying a cryptographic hash function $H[\cdot]$. In this way, we are not able to identify the hyperplanes that localize the data point and use information from multiple LSH keys to reconstruct the data point. We are, however, able to compare if two data points fall in the same localized region, by comparing their hashed keys.

2.3 Speech Recognition

Speech recognition is a type of pattern recognition problem, where the input is a stream of sampled and digitized speech data and the desired output is the sequence of words that were spoken. The pattern matching involves combining acoustic and language models to evaluate features which capture the spectral characteristics of the incoming speech signal. Most modern speech recognition systems use HMMs as the underlying model.

We view a speech signal as a sequence of piecewise stationary signals and an HMM forms a natural representation to output such a sequence of frames. We view the HMM that models the process underlying the observations as going through a number of states, in case of speech, the process can be thought of as the vocal tract of the speaker. We model each state of the HMM using a Gaussian mixture model (GMM) and correspondingly we can calculate the likelihood for an observed frame being generated by that model. The parameters of the HMM are trained over the labeled speech data using the Baum-Welch algorithm. We can use an HMM to model a phoneme which is the smallest segmental unit of sound which can provide meaningful distinction between two utterances. Alternatively, we can also use an HMM to model a complete word by itself or by concatenating the HMMs modeling the phonemes occurring in the word.

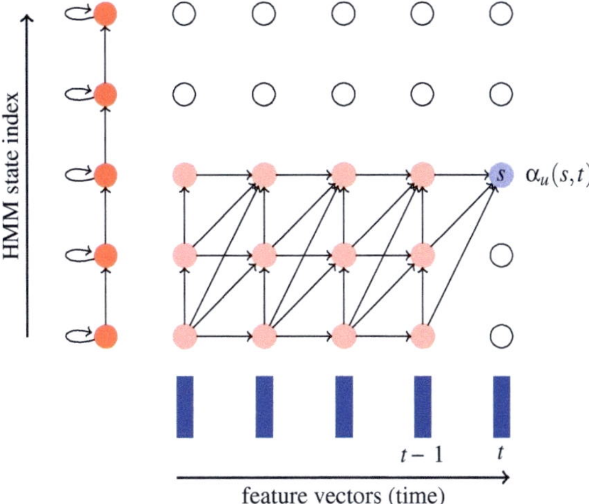

Fig. 2.6 A trellis showing all possible paths of an HMM while recognizing a sequence of frames

A useful way of visualizing speech recognition of an isolated word using an HMM is by a trellis shown in Fig. 2.6. Every edge in the graph represents a valid transition in the HMM over a single time step and every node represents the event of a particular observation frame being generated from a particular state. The probability of an HMM generating a complete sequence of frames can be efficiently computed using the forward algorithm. Similarly, the state sequence having the maximum probability for an observation sequence can be found using the Viterbi algorithm. In order to perform isolated word recognition, we train an HMM for each word in the dictionary as a template. Given a new utterance, we match it to each of the HMMs and choose the one with the highest probability as the recognized word.

References

Bimbot F, Bonastre J-F, Fredouille C, Gravier G, Magrin-Chagnolleau I, Meignier S, Merlin T, Ortega-Garcia J, Petrovska-Delacretaz D, Reynolds D (2004) A tutorial on text-independent speaker verification. EURASIP J Appl Signal Process 4:430–451
Campbell W (2002) Generalized linear discriminant sequence kernels for speaker recognition. In: IEEE international conference on acoustics, speech and signal processing
Campbell WM, Sturim DE, Reynolds DA, Solomonoff A (2006) SVM based speaker verification using a GMM supervector kernel and NAP variability compensation. In: IEEE international conference on acoustics, speech and signal processing
Carey M, Parris E, Bridle J (1991) Speaker verification system using alpha-nets. In: International conference on acoustics, speech and signal processing
Charikar M (2002) Similarity estimation techniques from rounding algorithms. In: ACM symposium on theory of computing

Datar M, Immorlica N, Indyk P, Mirrokni VS (2004) Locality-sensitive hashing scheme based on p-stable distributions. In: ACM symposium on computational geometry, pp 253–262

Davis S, Mermelstein P (1980) Comparison of parametric representations for monosyllabic word recognition in continuously spoken sentences. IEEE Trans Acoust Speech Signal Process ASSP 28(4):357

Dehak N, Kenny P, Dehak R, Dumouchel P, Ouellet P (2011) Front-end factor analysis for speaker verification. IEEE Trans Audio Speech Lang Process 19(4):788–798

Dunn RB, Reynolds DA, Quatieri TF (2000) Approaches to speaker detection and tracking in conversational speech. Digit Signal Process 10:93–112

Heck LP, Weintraub M (1997) Handset-dependent background models for robust text-independen speaker recognition. In: International conference on acoustics, speech and signal processing, vol 2. pp 1071–1074

Indyk P, Motwani R (1998) Approximate nearest neighbors: towards removing the curse of dimensionality. In: Proceedings of the ACM symposium on theory of computing, pp 604–613

Kenny P, Dumouchel P (2004) Experiments in speaker verification using factor analysis likelihood ratios. In: Odyssey, pp 219–226

Kulis B, Grauman K (2009) Kernelized locality-sensitive hashing for scalable image search. In: IEEE international conference on computer vision, pp 2130–2137

Mariéthoz J, Bengio S, Grandvalet Y (2009) Kernel-Based Text-Independent Speaker Verification, in Automatic Speech and Speaker Recognition: Large Margin and Kernel Methods (eds) J. Keshetand S. Bengio, John Wiley & Sons, Ltd, Chichester, UK. doi: 10.1002/9780470742044. ch12

Matsui T, Furui S (1995) Likelihood normalization for speaker verification using a phoneme- and speaker-independent model. Speech Commun 17(1–2):109–116

Reynolds DA (1997) Comparison of background normalization methods for text-independent speaker verification. In: European conference on speech communication and technology, vol 2. pp 963–966

Reynolds DA, Quatieri TF, Dunn RB (2000) Speaker verification using adapted Gaussian mixture models. Digit Signal Process 10:19–41

Rosenberg AE, Parthasarathy S (1996) Speaker background models for connected digit password speaker verification. In: International conference on acoustics, speech and signal processing

Chapter 3
Privacy Background

In this chapter, we introduce the basic concepts about privacy and privacy-preserving methodologies. We also introduce a formal model of privacy-preserving computation: secure multiparty computation (SMC) that we use in the remainder of the thesis and discuss some of the common underlying techniques such as homomorphic encryption and cryptographic hash functions.

3.1 What is Privacy?

Informally, privacy can be considered to be a privilege that an individual has to prevent its information from being made public to another individual, group, or society at large. Privacy can be in many forms such as physical, informational, organizational, and intellectual. In this thesis we mainly investigate informational privacy, specifically, the privacy with respect to a person's speech.

3.1.1 Definitions

As privacy is connected to every level of the human experience, it is an issue of considerable importance to society at large. The modern right to privacy is attributed to Warren and Brandeis (1890), who considered it as protecting "the inviolate personality" of an individual. This was in response to the incursions on personal privacy made possible due to the newly adopted technologies of that era such as photography and faster newspaper printing. In the last few decades, the importance of privacy has been further compounded due to the ubiquity and prevalence of technologies that allow individuals to interact and collaborate with each other. Privacy issues, therefore, have increasingly been at the focus of law makers, regulatory bodies, and consumers alike.

M. A. Pathak, *Privacy-Preserving Machine Learning for Speech Processing*,
Springer Theses, DOI: 10.1007/978-1-4614-4639-2_3,
© Springer Science+Business Media New York 2013

Although privacy is considered to be a basic human right, what constitutes as reasonable amount of privacy varies widely according to cultural and other socio-political contexts. This makes the task of precisely defining privacy fairly difficult (DeCew 2008). We, however, reproduce some of the commonly referred to definitions of privacy.

1. Privacy as control over information (Westin 1967).

   "Privacy is the claim of individuals, groups or institutions to determine for themselves when, how, and to what extent information about them is communicated to others."
2. Privacy as limited access to self (Gavison 1980).

   "A loss of privacy occurs as others obtain information about an individual, pay attention to him, or gain access to him. These three elements of secrecy, anonymity, and solitude are distinct and independent, but interrelated, and the complex concept of privacy is richer than any definition centered around only one of them."
3. Privacy as contextual integrity (Barth et al. 2006).

   "Privacy as a right to appropriate flows of personal information."

3.1.2 Related Concepts

Privacy is closely related to and alternatively used with the concepts of *anonymity* and *security* in common parlance. We briefly discuss the differences between these concepts below.

1. Privacy and Anonymity. Anonymity is the ability of an individual to prevent its identity from being disclosed to the public. Despite being closely related, privacy and anonymity are often separate goals. In privacy, the individual is interested in protecting his/her data, while in anonymity, the individual is interested in protecting only its identity. For instance, a web browsing user can potentially achieve anonymity by using an anonymous routing network such as Tor (Dingledine et al. 2004) to hide its IP address and location from a website, but the website provider will still have access to all the information provided by the user. In many situations, anonymity does not lead to privacy, as the information collected by the service provider can be correlated to identify individual users. An example of this was the AOL search data leak (Barbaro and Zeller 2006), where individual users were identified from anonymized search engine query logs. Another similar example is the de-anonymization of the Netflix dataset (Narayanan and Shmatikov 2008), where users were identified from anonymized movie rating data using publicly available auxiliary information.
2. Privacy and Security. In the context of information, security is the ability of protecting information and systems from unauthorized access. The unauthorized access can be in the form of public disclosure, disruption, deletion of the underlying information. Similarly, privacy and security differ in their intended goals. In a

multiparty system, security aims to protect the information from the unauthorized access of an eavesdropper who is not one of the participants, while privacy aims to protect the information provided by a party from the other parties. For instance, in a speech-based authentication system deployed in a client-server model, the speech input provided by the client is encrypted using the server's public key to protect against a third party that can intercept the communication. This, however, does not protect the privacy of the client's input from the server, as the latter needs to observe the input in order to do the necessary processing.

3.1.3 Privacy-Preserving Applications

We live in a world where there are vast amounts of data available from sources such as social media, medical services, and sensor networks. There is a wealth of information that can potentially be gained by applying data processing and machine learning techniques such as classification, clustering, and probabilistic inference methods to this data. In many situations, however, this data is often distributed among multiple parties in the form of social network users, the patient and multiple health care providers in the case of medical services, and sensor nodes controlled by different entities in sensor networks. In each of these cases, there are a combination of personal, legal, ethical, or strategic constraints against sharing the data among the parties that is required for performing the necessary computation. We discuss privacy-preserving computation in the context of three applications: biometric authentication, surveillance, and external third-party services.

1. Privacy in biometric authentication.
   Biometrics like voice, face, fingerprints, and other personal traits are widely used as robust features to identify individuals in authentication systems. Biometrics are inherently non-exact in their measurement, e.g., multiple recordings of a person speaking the same content would always be slightly different. We therefore cannot perform biometric matching by direct comparison, and require machine learning algorithms to utilize biometrics with high accuracy. There are, however, stringent personal privacy and legal constraints dealing with how biometrics are stored and used as they represent data derived from personal attributes. It is important to keep the biometric data secure to protect the privacy of users, and we require privacy-preserving machine learning algorithms that can perform the authentication using the secure data.
2. Privacy in surveillance.
   Machine learning algorithms are also used in *surveillance* applications such as speaker identification from wiretapped phone recordings, text classification over monitored emails, and object recognition from closed-circuit camera images. Although some form of surveillance is necessary to identify credible security threats, it has the unavoidable side-effect of infringing on the personal privacy of the innocent citizens. In a surveillance setting, privacy-preserving machine

learning algorithms can be used to aid investigating agencies to obtain necessary information without violating privacy constraints.

3. Privacy in Third-Party Services.

Privacy issues becomes relevant when an individual entity is interested in analyzing its data but does not have access to the required computational resources or algorithm implementations. An example of this would be an independent hospital interested in applying a classification model to calculate the likelihood of a disease from their patient records, or a small movie rental service interested in developing a recommendation algorithm based on reviews from its users. An increasingly useful option is to rely on a *cloud-based service* provider that has abundant computing power and sophisticated proprietary algorithms. Due to the same privacy and confidentiality constraints, many entities are reluctant to export their private data outside their control and are not able to utilize cloud-based services. Privacy-preserving computation aims to provide a framework for these types of computation without violating the privacy constraints of the participants.

3.1.4 Privacy-Preserving Computation in this Thesis

The central theme of this thesis is the design and implementation of the privacy-preserving computation frameworks similar to the ones mentioned above applied to speech processing. We mainly focus on privacy-preserving solutions for the speaker verification, speaker identification and speech recognition tasks as examples of biometric authentication, surveillance, and cloud-based service applications.

To create such privacy-preserving mechanisms, we use cryptographic primitives such as homomorphic encryption i.e., schemes in which we can perform operations on unencrypted data by performing corresponding operations on encrypted data. Rather than merely combining such primitives to build machine learning algorithms, we aim to find the synergy between the primitives, often in the form of a trade-off between privacy, accuracy, and efficiency. In some cases, we need to modify the machine learning algorithm itself to fit the right kind of cryptographic primitive and vice versa, in order to improve the scalability of the privacy-preserving solution.

3.2 Secure Multiparty Computation

Secure multiparty computation (SMC) is a sub-field of cryptography that concerns with settings in which n parties are interested in jointly computing a function using their input data while maintaining the privacy of their inputs. We overview the basic concepts of SMC in this section.[1]

---

[1] Note on presentation: many cryptography textbooks, such as (Goldreich 2001, 2004), first begin with cryptographic primitives and definitions based on computational complexity theory, and then

The privacy in SMC protocols is established by the constraint that no party should learn anything apart from the output and the intermediate steps of the computation. We consider that the parties $p_1, \ldots, p_n$ respectively have inputs $x_1, \ldots, x_n$ and we need to compute the function with n outputs, $f(x_1, \ldots, x_n) = (y_1, \ldots, y_n)$. The privacy constraints require that the party p_i provides the input x_i and obtains the output y_i, and does not observe the input or output belonging to any other party.

In this model, the function of interest f could be any computable function. A useful example is the millionaire's problem (Yao 1982), where two millionaires interested in finding out who is richer, without disclosing their wealth. If the net worth of the two millionaires is \$$x$ million and \$$y$ million respectively, the underlying computation is the comparator function $f(x, y) = \mathbb{1}(x > y)$, i.e., the function has a single bit output 1 if $x > y$ and 0 if $x \leq y$. At the end of the computation, the output of the function is available to both the millionaires. The main privacy criteria here is that each millionaire does not learn anything else about the other millionaire's wealth from the computation apart from this information. We summarize the problem for two placeholder parties "Alice" and "Bob" below.

Private Comparison: Millionaire Problem.
Inputs:

(a) Alice has the input x.
(b) Bob has the input y.

Output: Alice and Bob know if $x > y$.

In speech-processing applications that we consider in this thesis, the two parties are client and server with the speech input and speech models respectively. The parties are interested in performing tasks such as speaker verification, speaker identification, and speech recognition. This usually involves computing a score of the speech models with respect to the speech input. In this section we use a running example to illustrate SMC protocols: we consider the client speech input to be a d-dimensional vector x represented using MFCC features and a linear classification model w, also a d-dimensional vector, such that the score is given by the inner product $w^T x$. The server will take an action depending on whether this score is above or below a pre-determined threshold θ. This model is simplistic, because, as we will see in the future chapters, most speech processing tasks require many other steps apart from inner product computation.

In this model, our the privacy criteria require that the input x should be protected from the server and the model w should be protected from the client. At the end of the computation, the server learns if $w^T x > \theta$, and the client does not learn anything. We summarize this problem below.

(Footnote 1 continued)
develop the theory of SMC protocols. We refer to this as the bottom-up approach. For the ease of exposition and maintaining the speech-based application theme of the thesis, we instead take a top-down approach of starting with SMC protocols for practical applications and work down to the necessary cryptographic primitives. The readers are also encouraged to refer to Cramer (1999), Cramer et al. (2009) for a detailed review of the subject.

Private Linear Classification.
Inputs:

(a) Client has the speech input vector x.
(b) Server has the classifier model w and the threshold θ.

Output: Client and Server know if $w^T x > \theta$.

It is important that no individual party gets access to the inner product $w^T x$, as it can be used to infer the input of the other party, e.g., the server can use w and $w^T x$ to gain more information about x, which it should not obtain otherwise. To solve the linear classifier problem, we therefore require two steps, private inner product computation and private comparison.

In the above discussion, we introduced SMC problem intuitively; we now begin outlining various assumptions that we make in creating SMC protocols and the types of adversarial behavior of the parties. In the remainder of the section, we will look at different cryptographic primitives and techniques that allow us to perform these steps with privacy.

3.2.1 Protocol Assumptions

The privacy requirements of any SMC protocol ultimately depends on the assumptions made on the parties and their adversarial behavior. We discuss some of the conditions below.

1. Setup. We assume that all parties have copies of the instruction sequence of the protocol and know when they are required to perform their actions. Also, in cases where we would need to use an encryption scheme, we assume the existence of a public-key infrastructure and the parties have access to the public keys belonging to other parties.
2. Communication channels. We consider a private-channel model where an adversary cannot listen in on messages sent between parties. However, in the general model, an adversary can tap any communication channels between parties and the channels by themselves do not provide any security. Similarly, we also assume that the channels are reliable, i.e., an adversary cannot modify or delete any message sent between parties and also cannot pretend to be another party and receive the messages sent to it. Using appropriate primitives such as public-key encryption schemes, private-channel model can be simulated using open channels.
3. Computational limitations. By default we assume the adversaries to be computationally bounded in probabilistic polynomial time. This requirement is important because a computationally unbounded adversary would be able to break through the cryptographic primitives and we would not be able to protect the privacy of other participants' inputs.

3.2.2 Adversarial Behavior

As discussed above, the motivation for parties to use SMC protocols is to perform private computation while protecting the privacy of the inputs. A basic question is: whom are the parties trying to protect their privacy against? If we assume that the parties are entirely honest, i.e., are uninterested in learning anything about the input provided by other parties, we could simply require all parties to share their data with each other and perform the necessary computation collaboratively and later agree to delete the data they had obtained. In practice, however, this model is trivial, as parties would not be comfortable in sharing their input with other parties, e.g., in the private linear classification problem, the client would not want to disclose its input to the server due to confidentiality and the server would not want to disclose the parameters of its classifier model as may contain proprietary information.

In a more realistic setting, we assume the parties to be *semi-honest* or *honest-but-curious*. In this model a party only tries to gather as much information as possible about the other party's input from the intermediate steps of the protocol, but does not try to actively disrupt the protocol, e.g., by providing malformed inputs. The parties are able to do this by maintaining a log of all actions and information that the party obtained during each step of the protocol. Due to this behavior, the trivial model of the parties sharing their data with each other would lead to a loss of privacy and we require additional primitives to protect the privacy criteria. Although the semi-honest model suffices in some cases, we also consider a *malicious* party with unconstrained adversarial behavior, in which the party would also disrupt the protocol using all possible means in addition to semi-honest behavior. A common malicious behavior is to pollute the output of the computation by submitting malformed input.

Apart from the behavior of individual parties, it is also common to consider an *adversary* who gains control over and corrupts a subset of the parties. The adversary is able to gain complete access to the entire history of the party including the data inputs and the logs stored by the party. There are many variations to the nature of the adversary, we discuss a few types below.

1. Monolithic Adversary. There is a single adversary that controls a subset of k parties (out of n) and causes them to behave maliciously by modifying the steps of the protocol and coordinates their actions.
2. Passive and Active Adversary. A passive adversary gains access only to the internal state of party and cannot modify the operations of the party. The adversary, however, utilizes that information to learn more about other parties. This behavior causes the affected parties to have semi-honest behavior. An active adversary, on the other hand, fully controls the party and causes it to perform malicious actions.
3. Static and Adaptive Adversary. In case of a static adversary, we assume that the set of dishonest parties is fixed before the protocol execution is started, but it is not known to the honest parties. Alternatively, an adversary is considered to be adaptive when it can choose which parties to corrupt while the protocol is being executed and makes its decision based on the partial information obtained from the intermediate steps of the protocol.

4. Threshold Adversary. There are limits to how many parties can be corrupted by
   the adversary, as in the worst case, it does not make sense to consider the privacy of
   a protocol where all parties are corrupted. In many cases, we consider a threshold
   adversary, that can corrupt any subset of t parties (out of n).

3.2.3 Privacy Definitions: Ideal Model and Real Model

One fundamental question is how do we know that an SMC protocol actually does
perform the desired computation while satisfying the necessary privacy constraints?
The classical approach of defining privacy is to write down a list of the privacy
requirements of the protocol and manually make sure that they are satisfied. The
problem with this approach is that the requirements of non-trivial protocols are hard
to formalize and are often endless; it is difficult to determine if we have covered
all in our set. A completely different approach towards this problem is to consider
the protocols in two worlds: *ideal model* and *real model*. In the ideal model, we
provide a specification of the desired computation of the protocol and in the real
model, we consider the protocol with adversarial behavior. We consider a protocol
to satisfy privacy if the real model behavior cannot be distinguished from the ideal
model behavior.

In the ideal model, we consider a trusted third party (TTP) and all parties are able
to communicate with it securely. The parties submit their inputs to the TTP, which
performs the necessary computation and submits the output back to the respective
parties. The TTP is expected to be incorruptible: it executes a pre-specified set of
commands on the input data provided by the parties, correctly outputs the result and
later deletes the input data. We show this in Fig. 3.1. No party ever gets the opportunity
to observe the input or the output belonging to any other party, and therefore the the
privacy criteria of the SMC protocol are trivially satisfied in the ideal model.

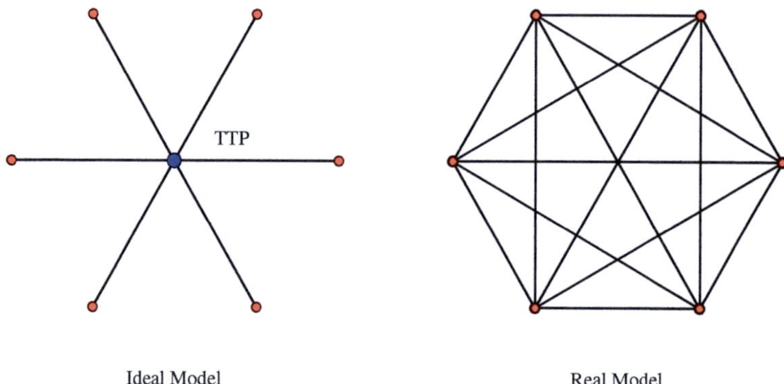

Fig. 3.1 An SMC protocol with ordinary participants denoted by *red* (*corner nodes*) emulating the
behavior of a trusted third party (TTP) denoted by *blue* (*center node*)

TTPs, however, do not exist in the real world and the goal of SMC protocol is to perform the same computation with untrusted parties in the presence of adversaries, after satisfying the same privacy constraints.

3.2.3.1 Theoretical Results

One of the most important result of SMC and theoretical cryptography was to prove that any function can be computed privately. In the information-theoretic model in the presence of an active and adaptive adversary, Ben-Or et al. (1988) establish that any function can be efficiently computed by n parties, if less than $\frac{n}{2}$, i.e., $\frac{n}{3}$ parties are corrupted. Similar results were also obtained by Chaum et al. (1988). The efficiency of these constructions was improved on by Gennaro et al. (1998), Beerliová-Trubíniová and Hirt (2008).

3.2.4 Encryption

Encryption is one of the most important operations in cryptography. This operation in the form of an encryption function $E[\cdot]$ allows one party to transform a message or data x, referred to as the *plaintext*, to an obfuscated form or *ciphertext $E[x]$*. Encryption is usually coupled with a corresponding decryption operation defined by the decryption function $E^{-1}[\cdot]$: the ciphertext $E[x]$ can then be converted back to the original plaintext x using a key. The encryption and decryption functions together are referred to as the *cryptosystem*.

Encryption has been widely used between communicating parties, archetypically referred to as "Alice" and "Bob", to maintain secrecy of communication against an adversary "Eve" who might be interested in eavesdropping in on the conversation. In a *symmetric-key* cryptosystem, only one key is used for both encryption and decryption operations. This key effectively is a shared secret between Alice and Bob. If Alice and Bob wish to communicate with each other, they would first need to securely set up the key. Examples of symmetric-key cryptosystems include DES, AES, Blowfish.

Conversely, an *asymmetric-key* or *public-key* cryptosystem provides a pair of public and private keys, where the public key is used for the encryption operation and private key is used for decryption operation. As the name suggests, the parties publish their public keys, while they keep the private key with themselves. The public key does not provide any information about the private key. The public-key cryptosystem allows Alice to send an encrypted message to Bob without requiring a secure initial key exchange between them. Alice can simply look up Bob's public key, encrypt her message using it, and send it to Bob. Bob can then decrypt this message using his decryption key. This arrangement finds widespread use in applications such as electronic commerce. Although, various forms of symmetric-key cryptosystems have existed since antiquity, a mechanism for exchanging keys was first proposed in the pioneering work of Diffie and Hellman (1976).

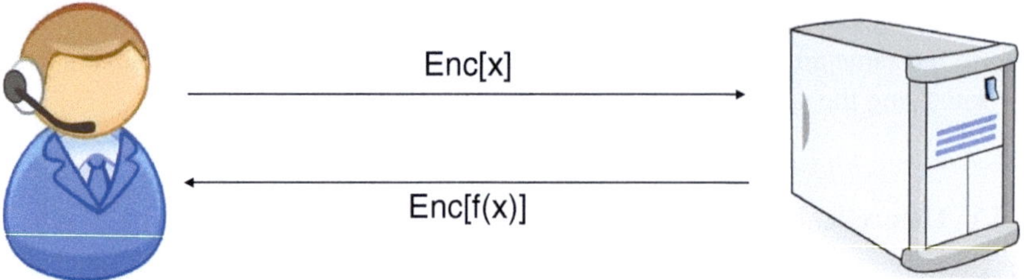

Fig. 3.2 Homomorphic encryption in a client server setting

3.2.4.1 Homomorphic Encryption

Homomorphic encryption (HE) schemes are a special type of cryptosystems that allow for operations to be performed on ciphertexts without requiring knowledge of the corresponding plaintexts. A homomorphic encryption scheme is typically a public-key cryptosystem. This allows for Alice to create a public-private key pair, encrypt her data, and send the ciphertexts to Bob after publishing the public key. Bob can perform the necessary computation directly on the ciphertexts without being able to decrypt them, as he does not have the key. This arrangement finds use in client-server settings, where Alice is a client user, and Bob is a server (Fig. 3.2). The concept of homomorphic encryption was first introduced by Rivest et al. (1978a) and it is one of the most important technique for creating SMC protocols in this thesis.

More formally, if two data instances x and y are encrypted to $E[x]$ and $E[y]$ using a homomorphic encryption scheme, we can obtain the encryption of the result of an operation $\otimes$ performed on x and y by performing another, possibly the same, operation $\oplus$ directly on the encrypted versions of x and y,

$$E[x \otimes y] = E[x] \oplus E[y].$$

A cryptosystem in which we can perform any operations on the plaintext by performing operations on corresponding ciphertext is called a *fully* homomorphic cryptosystem (FHE). The first such cryptosystem was proposed in the breakthrough work by Gentry (2009, 2010a). Although the construction satisfies the necessary properties for FHE, it is found to be computationally inefficient to be used in practice (Gentry 2010b) and developing practical FHE schemes is an active area of research (Lauter et al. 2011).

There are, however, well-established and efficient partially homomorphic encryption schemes that allow a limited set of operations to be performed on plaintext by performing operations on ciphertext, e.g., unpadded RSA (Rivest et al. 1978b) is multiplicatively homomorphic, El-Gamal (1985), Damgård and Jurik (2001), Benaloh (1994), Paillier (1999) are examples of additively homomorphic cryptosystems. Please refer to Fontaine and Galand (2007) for a survey.

We briefly overview the Paillier (1999) and BGN (Boneh et al. 2005) cryptosystems below. We use the homomorphic properties of these cryptosystems to construct higher-level primitives that are useful in creating a protocol for the private linear classifier problem introduced at the beginning of Sect. 3.2.

1. Paillier Cryptosystem

   Paillier cryptosystem is a public-key cryptosystem satisfies additive homomorphism. This cryptosystem is based on modular arithmetic with a finite field of order n^2, where n is a product of two large primes p and q. The primes p and q are chosen such that n is a 1024-bit number. Although it is very efficient to compute $n = pq$, there is currently no feasible algorithm known to factor n into p and q. The security of the Paillier cryptosystem is based on the decisional composite residuosity assumption (DCRA). We briefly overview this assumption by first defining the concept of residuosity below.

   Residue.
   Given a group $\mathbb{Z}_{n^2}$, a number z is the ith residue mod n^2 if there exists y such that

   $$z = y^i \mod n^2.$$

   DCRA deals with checking for nth residuosity. We state this assumption below.

   Decisional Composite Residuosity Assumption.
   For a given a number z, it is hard to check if it is an n-th residue mod n^2, i.e., if

   $$z = y^n \mod n^2.$$

   We use the term *hardness* to mean computational infeasibility, which is equivalent to saying that there is no known efficient algorithm to solve the decisional composite residuosity problem.

   We now describe the construction of the Paillier cryptosystem. As discussed above, there are three components of the cryptosystem: a key generation algorithm that outputs a public-private key pair, an encryption function, and a decryption function.

   Notation: $\mathbb{Z}_{n^2}^* \subset \mathbb{Z}_{n^2} = \{0, \ldots, n^2 - 1\}$ denotes the set of non-negative integers that have multiplicative inverses modulo n^2, $\lambda = \text{lcm}(p - 1, q - 1)$ is the Carmichael function, and $L(x) = \frac{x-1}{n}$.

   a. Key Generation.
      i. Choose two large random primes p and q independently, and let $n = pq$.
      ii. Choose a random number $g \in \mathbb{Z}_{n^2}^*$ such that $\gcd(L(g^\lambda \mod n^2), n) = 1$. (n, g) is the public key, and (p, q) is the private key.
   b. Encryption. For a plaintext $m \in \mathbb{Z}_N$, the ciphertext is given by

   $$E[m] = g^m \, r^n \mod n^2, \tag{3.1}$$

   where $r \in \mathbb{Z}_n^*$ is chosen randomly.
   c. Decryption. For a ciphertext $c \in \mathbb{Z}_{n^2}$, the corresponding plaintext is given by

$$E^{-1}[c] = \frac{L(c^\lambda \quad \mathrm{mod}\ n^2)}{L(g^\lambda \quad \mathrm{mod}\ n^2)}. \tag{3.2}$$

Paillier (1999) show that the decryption function satisfies the property that $E^{-1}[E[m]] = m \mod n$.

We choose a different random number r during each application of the encryption function, but the decryption function works irrespective of the value of r used. We discuss the importance of this at the end of this subsection.

We can easily verify that the Paillier cryptosystem satisfies additive homomorphism by multiplication, i.e., we can multiply two ciphertexts to add their corresponding plaintexts.

$$\begin{aligned}
E[m_1]E[m_2] &= g^{m_1}\, r_1^n\, g^{m_2}\, r_2^n \quad \mathrm{mod}\ n^2 \\
&= g^{m_1+m_2}\, (r_1 r_2)^n \quad \mathrm{mod}\ n^2 \\
&= E[m_1 + m_2].
\end{aligned} \tag{3.3}$$

The Paillier cryptosystem, therefore, homomorphically maps the plaintext group with addition $(\mathbb{Z}, +)$ to ciphertext group with multiplication, $(\mathbb{Z}, \cdot)$.

As a corollary, we can also multiply a ciphertext by a plaintext.

$$\begin{aligned}
[E[m_1]]^{m_2} &= [g^{m_1}\, r_1^n \quad \mathrm{mod}\ n^2]^{m_2} \\
&= g^{m_1 m_2}\, (r_1^{m_2})^n \quad \mathrm{mod}\ n^2 \\
&= E[m_1 m_2].
\end{aligned} \tag{3.4}$$

These homomorphic properties allow us to compute the inner product of an encrypted vector with another vector. We construct such a protocol below. In the private linear classifier problem, this protocol can be used by the server to compute the inner product between the client's input vector x and its weight vector w.

Private Inner Product Protocol.
Inputs:

(a) Client has a d-dimensional vector $x = (x_1, \ldots, x_d)$.
(b) Server has a d-dimensional vector $w = (w_1, \ldots, w_d)$.

Output: Server obtains the encrypted inner product $E[w^T x]$.

a. Client generates a public-private key pair for the Paillier cryptosystem with encryption and decryption functions $E[\cdot]$ and $E^{-1}[\cdot]$. The client sends the encryption function to the server.
b. Client encrypts its input vector to obtain $E[x] = (E[x_1], \ldots, E[x_d])$. Client sends the encrypted vector to the server.
c. Server performs element-wise homomorphic multiplication with the encrypted vector and its input vector, which it has in plaintext.

$$(E[x_1]^{w_1}, \ldots, E[x_d]^{w_d}) = (E[x_1w_1], \ldots, E[x_dw_d]).$$

d. Server homomorphically adds these elements to obtain the encrypted inner product.

$$\prod_i E[x_iw_i] = E\left[\sum_i x_iw_i\right] = E[w^Tx].$$

The server only has access to the client's public key and not the private key. It observes the client's input vector in ciphertext and cannot decrypt it, but is still able to compute the inner product.

2. BGN Cryptosystem

The BGN (Boneh-Goh-Nissim) cryptosystem is designed to satisfy additive homomorphism like the Paillier cryptosystem, but in addition to that, it supports one homomorphic multiplication operation on ciphertext. Being able to perform either addition or multiplication alone, or in other words, being able to evaluate 2-DNF formulas on ciphertexts has its limitations; given two encrypted vectors $E[x] = (E[x_1], \ldots, E[x_m])$ and $E[y] = (E[y_1], \ldots, E[y_m])$, we are not able to compute the inner product $E[x^Ty]$ without requiring an interactive protocol involving the party with the private key.

As compared to the Paillier cryptosystem, the BGN cryptosystem is constructed using groups defined over elliptic curves (Miller 1985; Koblitz 1987). The cryptosystem relies on the the subgroup decision problem (SDP) as its underlying basis for security. Given a group $\mathbb{G}$ of an composite order n, i.e., n is a product of two primes q_1 and q_2 and a subgroup $\mathbb{G}'$ of $\mathbb{G}$. By the Lagrange theorem, the order of $\mathbb{G}'$ divides n, which implies that the order of $\mathbb{G}'$ is either q_1 or q_2. It is computationally difficult to determine such a subgroup given the hardness of factorizing $n = q_1q_2$. SDP deals with checking the membership of an element in the subgroup. We define the corresponding assumption below.

Subgroup Decision Assumption.
Given a group $\mathbb{G}$ of order $n = q_1q_2$, where q_1 and q_2 are large primes, and an element $x \in \mathbb{G}$, it is hard to check if x is also an element of a subgroup of $\mathbb{G}$ with order q_1.

The BGN key generation algorithm produces a tuple $(q_1, q_2, \mathbb{G}, \mathbb{G}_1, e)$ where q_1 and q_2 are τ-bit prime numbers, $\mathbb{G}$ and $\mathbb{G}_1$ are groups, both of order $n = q_1q_2$ defined over the points on an elliptic curve and $e : \mathbb{G} \times \mathbb{G} \mapsto \mathbb{G}_1$ is a bilinear map. We choose two random generators g, u of the group $\mathbb{G}$ and set $h = u^{q_2}$, thereby making it the generator of a subgroup of $\mathbb{G}$ of order q_1. The public key pk is given by the tuple $(n, \mathbb{G}, \mathbb{G}_1, e, g, h)$ and the private key sk is the number q_1.
For the plaintexts in the set $\mathscr{D} = \{1, \ldots, T\}$, we define the encryption function $E : \mathscr{D} \mapsto \mathbb{G}$ as $E[x] = g^xh^r$, where r is a randomly chosen number in $\mathbb{Z}_n = \{0, \ldots, n-1\}$ and the exponentiation is modular in $\mathbb{G}$. Using a different value of r in each execution of the encryption function provides semantic security—two different encryptions of a number x say, $E[x; r_1]$ and $E[x; r_2]$ will have different values but

decrypting each of them will result in the same number x. We define the decryption function $E^{-1} : \mathbb{G} \mapsto \mathcal{D}$ as $E^{-1}[c] = \log c^{q_1}$, where the discrete-log is computed to the base g^{q_1}. Although discrete-log is considered to be a computationally hard problem, here we compute it only in the range of the plaintext $\mathcal{D}$ which is much smaller than n. The discrete-log over plaintext space can be computed in expected-time $\tilde{O}(\sqrt{T})$ using Pollard's lambda method (Menezes et al. 1996).

We can easily verify that the encryption function is additively homomorphic; for any two ciphertexts $E[x]$ and $E[y]$,

$$E[x]\,E[y] = g^x h^{r_1} g^y h^{r_2} = g^{x+y} h^{r_1+r_2} = E[x+y].$$

As a corollary, for any ciphertext $E[x]$ and plaintext y, $E[x]^y = E[x\,y]$. We obtain multiplicative homomorphism by applying the bilinear mapping e; for any two ciphertexts $E[x]$ and $E[y]$, it can be shown that

$$e(E[x], E[y])h_1^r = g_1^{xy} h_1^{r'} = E_{\mathbb{G}_1}[xy],$$

where $g_1 = e(g, g)$, $h_1 = e(g, h)$, and $r, r' \in \mathbb{Z}_n$. By this operation, the ciphertexts from $\mathbb{G}$ are mapped into $\mathbb{G}_1$, and the cryptosystem is still additively homomorphic in that space,

$$E_{\mathbb{G}_1}[x_1]\,E_{\mathbb{G}_1}[x_2] = E_{\mathbb{G}_1}[x_1 + x_2],$$

but we cannot perform further multiplications on these ciphertexts. We hereafter omit the subscript from $E_{\mathbb{G}_1}$ for conciseness.

As mentioned above, one motivation of using BGN is to perform homomorphic multiplication. This allows us to create a protocol for computing the inner product of two encrypted vectors. This protocol is useful when the client is interested in protecting the classifier weight vector from the server, e.g., in case of authentication systems such as speaker verification.

Private Inner Product Protocol over Ciphertexts.
Inputs:

(a) Client has a d-dimensional vector $x = (x_1, \ldots, x_d)$ and both encryption and decryption functions $E[\cdot]$ and $E^{-1}[\cdot]$.
(b) Server has an encrypted d-dimensional vector $E[w] = (E[w_1], \ldots, E[w_d])$ and the encryption function $E[\cdot]$.

Output: Server obtains the encrypted inner product $E[w^T x]$.

a. Client encrypts its input vector to obtain $E[x] = (E[x_1], \ldots, E[x_d])$. Client sends the encrypted vector to the server.
b. Server performs element-wise homomorphic multiplication with the encrypted vector and its input vector, which it has in plaintext.

$$(e(E[x_1], E[w_1]), \ldots, e(E[x_d], E[w_d])) = (E[x_1 w_1], \ldots, E[x_d w_d]).$$

c. Server homomorphically adds these elements to obtain the encrypted inner product.

$$\prod_i E[x_i w_i] = E\left[\sum_i x_i w_i\right] = E[w^T x].$$

The server only has access to encryption function and cannot observe the client input or even its own input in plaintext. The server is still able to compute the inner product.

3.2.4.2 Semantic Security

In the encryption functions of both Paillier and BGN cryptosystems, we considered a random number r. By using a new random number in each encryption function application, we will get different ciphertexts $c_1 = E_{r_1}[x]$ or $c_2 = E_{r_2}[x]$ for the same plaintext x. The cryptosystems have the property that both these ciphertexts would decrypt to the same plaintext x, i.e., $E^{-1}[c_1] = E^{-1}[c_2] = x$. This is referred to as *probabilistic encryption* (Goldwasser and Micali 1984).

This property is very useful; if we are encrypting a data with a small number of unique elements, e.g., binary data with 0 and 1, in absence of such an encryption function, an adversary can statistically determine the proportion of the inputs and gain significant information about the data. More rigorously, we require an adversary to not learn anything from observing the ciphertext alone. This definition of privacy is called *semantic security* (Goldwasser and Micali 1982). A semantically secure public-key cryptosystem ensures that it is infeasible for a computationally bounded adversary to use the ciphertext and the public key to gain information about the plaintext.

Semantic security provides protection against chosen plaintext attacks (CPA). In this model, the adversary generates ciphertexts from arbitrarily chosen plaintexts using the public key, and attempts to gain information about the user, typically by trying to reconstruct the private key. Due to the requirements of semantic security as discussed above, the adversary is not able to do so.

One limitation of semantic security is that it does not protect against chosen ciphertext attacks (CCA). In this model, the adversary can obtain the decrypted plaintexts from the user for any ciphertext of its choosing.

3.2.5 Masking

In the previous subsection, we considered practical homomorphic cryptosystems that allow only a subset of operations that can be performed on ciphertexts, as opposed to fully homomorphic encryption schemes that are currently impractical. One question is how do we privately compute a function that requires operations beyond those provided by the cryptosystem? e.g., if we need to multiply two ciphertexts encrypted using the Paillier cryptosystem.

A solution for this problem is called *blinding* or *masking*, where one party modifies the input using a random number in order to obfuscate from another party. Suppose Bob has the number x, and is interested in hiding it from Alice. In *additive masking*, Bob can generate a random number r and transfer $x + r$ to Alice. This masking is information theoretically secure, if r is randomly chosen from the set of plaintexts, i.e., the domain of encryption function $\mathbb{Z}_n$, Alice cannot gain any information about x from $x + r$. This is because if r can take n values (256-bit or 1024-bit), then $x + r$ will take also n values, implying that the entropy of $x + r$ is equal to r. Additive masking is equivalent to *additive secret sharing*, as Alice and Bob have shares $x + r$ and $-r$ respectively, that add up to x. For Alice, there will be n ways to split $x + r$, which would be infeasible in practice.

Additive masking can also be performed over ciphertexts using homomorphic encryption. If Bob has a ciphertext $E[x]$ encrypted using Alice's public-private key pair, he can still add the random number r to $E[x]$ by first encrypting it using Alice's public key and computing $E[x]E[r] = E[x + r]$. After Bob sends this ciphertext to Alice, she can decrypt it to obtain $x + r$, which is additively masked. Even in this case, Alice and Bob would have additive shares of x.

Another form of masking is *multiplicative masking*, where Bob can multiply his input x by a random number q to obtain qx in order to hide it from Alice. Multiplicative masking hides less information than additive masking even after choosing q randomly from the set of plaintexts. This is because the range of qx is not equal to the range of q, e.g., if x is an even number, qx will always be an even number, which is not the case for additive masking. Information theoretically, there will be $\log n$ ways to split qx, implying that multiplicative masking is logarithmically weaker than additive masking.

3.2.5.1 Interactive Protocols Using Masking

As we mentioned at the beginning of this subsection, we are interested in augmenting the homomorphic operations of a cryptosystem with masking in an interactive protocol. As an example, we construct a protocol for performing homomorphic multiplication of ciphertexts using only additively homomorphic Paillier cryptosystem.

Private Multiplication Protocol.
Inputs:

(a) Alice has a and and both encryption and decryption functions $E[\cdot]$ and $E^{-1}[\cdot]$.
(b) Bob has $E[b]$ and the encryption function $E[\cdot]$.

Output: Bob obtains $E[ab]$.

1. Bob generates a random number r and encrypts it to obtain $E[r]$. Bob homomorphically adds it to his input to obtain $E[b + r]$ which he sends to Alice.
2. Alice decrypts the ciphertext $E[b + r]$ to obtain $b + r$. She multiplies this with a to obtain $ab + ar$.

3. Alice encrypts the product and also her input a, and she sends the ciphertexts $E[ab + ar]$ and $E[a]$ to Bob.
4. Bob homomorphically multiplies the plaintext r with $E[a]$ to obtain $E[ar]$ and homomorphically subtracts it from the product to obtain $E[ab]$.

Bob is able to homomorphically multiply ciphertexts with this protocol, something which is not possible using the homomorphic operations of Paillier cryptosystem alone. This protocol, however, requires 3 encryption and 1 decryption operation and the transfer of 3 ciphertexts. This is more expensive than using the BGN cryptosystem which can directly perform one homomorphic multiplication on ciphertexts.

As another example, we construct a protocol for the private linear classification problem introduced at the beginning of Sect. 3.2, using additive masking and the inner product protocols discussed above. We can either use the Paillier or the BGN cryptosystem for this purpose.

Private Linear Classification Protocol.
Inputs:

(a) Client has the speech input vector x.
(b) Server has the classifier model w and the threshold θ.

Output: Client and Server know if $w^T x > \theta$.

1. Client generates a public-private key pair with encryption and decryption functions $E[\cdot]$ and $E^{-1}[\cdot]$. The client sends the encryption function to the server.
2. Client and server execute the private inner product protocol, and as an output the server obtains $E[w^T x]$.
3. Server generates a random number r and homomorphically adds it to the inner product to get $E[w^T x + r]$. Server transfers the masked inner product to the client.
4. Client decrypts the server's message to obtain $w^T x + r$.
5. Client and the server execute the private comparison protocol using inputs $w^T x + r$ and $\theta + r$. As output the client and server know if $w^T x > \theta$.

The server masks the inner product $w^T x$, because the client can use its input x to gain information about w. In Step 5, client and the server hold additive shares of $w^T x$. The server uses the masked value of the threshold θ as input to the private comparison protocol, which is equivalent to the client and the server using $w^T x$ and θ respectively as input.

3.2.6 Zero-Knowledge Proofs and Threshold Cryptosystems

In most of the discussion above, we considered SMC protocols for *semi-honest* parties, i.e., we assume the parties to follow all the steps of the protocol and not use spurious inputs. Although the semi-honest assumption is valid in many settings, there are cases where the parties can have *malicious* behavior. Common examples of this include authentication systems, where the output of the SMC protocol could be sensitive. In order to construct protocols that can satisfy the privacy constraints, we need additional cryptographic primitives such as zero-knowledge proofs and threshold cryptosystems. In this subsection, we present a brief overview of these concepts, with an emphasis on creating SMC protocols that protect against malicious adversaries. Zero-knowledge proof (ZKP) is a deep area of theoretical cryptography with many applications beyond SMC protocols; please refer to Chap. 4 of Goldreich (2001) for a detailed review.

Goldwasser et al. (1985) originally introduced ZKP as a mechanism for one party to establish to another party that a statement is true without giving away additional information about that statement. We refer to the party establishing the proof as the *prover* and the other party as the *verifier*. Quisquater et al. (1989) present a classic example of ZKPs, where there is a cave with a door that can be opened with a secret spell, and a prover Peggy attempts to give a proof to the verifier Victor that she knows the spell. The ZKP criteria require that Victor should be convinced about Peggy's knowledge of the spell without learning what the spell is.

A valid ZKP satisfies three properties:

1. Completeness. If the statement is true, the verifier will always accept the proof.
2. Soundness. If the statement is false, the verifier will accept with only a negligible probability.
3. Zero Knowledge. The verifier does not learn anything about the statement apart from its veracity.

There are two types of ZKPs: interactive and non-interactive. The proofs we discussed above, e.g., Quisquater et al. (1989), are interactive. Another variant of ZKP is non-interactive proofs (Blum et al. 1988), in which a prover can establish a proof by submitting only a reference string to the verifier without requiring further interaction from the prover.

In their breakthrough work, Cramer et al. (2001) make a connection between ZKPs and homomorphic encryption. Firstly, they propose threshold homomorphic cryptosystems, which in addition to satisfying the properties of a regular homomorphic cryptosystem, provide a different model for decryption. In a threshold cryptosystem, there is only one public key pk, but the decryption key is split among multiple parties in the form of shares $sk_1, \ldots, sk_m$, typically obtained using a secret sharing protocol. An individual party can encrypt a plaintext message x to obtain the ciphertext $E[x]$, but a single share of the private key cannot be used to perform the decryption. A party can only obtain a share of the plaintext s_i after applying the decryption function using the share of the key: $E_{sk_i}^{-1}[E[x]] = s_i$, with the property that the share s_i alone does not provide information about the plaintext x. A majority of the parties need to execute a decryption protocol together to obtain the complete decryption result.

Secondly, the threshold homomorphic cryptosystem is augmented with two non-interactive ZKPs:

1. Proof of Knowledge of Plaintext (POK). A party submitting a ciphertext $c = E[x]$ can provide a ZKP $POK(E[x])$ that is a proof that the party has access to the corresponding plaintext x. In a cryptosystem providing semantic security, using

encryption function with different randomization r_1 and r_2 will result in two ciphertexts $c_1 = E[x; r_1]$ and $c_2 = E[x; r_2]$. Using *POK*, a party can establish that they are the same without disclosing the plaintext x. This is useful when the party is participating in two execution of the protocol with the same input.

2. Proof of Correct Multiplication (POCM). A party that is multiplying a ciphertext $c = E[x]$ with a plaintext a homomorphically to obtain $E[ax]$, can provide a ZKP $POCM(E[x], E[a], E[ax])$. Using this proof, the verifier can check if the ciphertext $E[x]$ was indeed correctly multiplied by a, without observing either x or a. We typically use this proof to check if a party is performing the steps of the protocol correctly.

Damgård and Jurik (2001) extend the Paillier cryptosystem to provide threshold decryption and the ZKPs discussed above. Kantarcioglu and Kardes (2008) use this as primitives to construct SMC protocols for malicious adversaries for operations such as computing the inner product. ZKPs and threshold cryptosystem impose a significant computational and communication overhead as compared to the protocols using conventional homomorphic cryptosystems. The overhead in the private inner product of Kantarcioglu and Kardes (2008) is about 700 times as compared to the private inner product protocol we constructed in Sect. 3.2.4 using Paillier encryption.

3.2.7 Oblivious Transfer

Oblivious transfer (OT) is another useful primitive introduced by Rabin (1981). The motivation of this work was to construct a mechanism for two parties to exchange secrets without a trusted third party or a safe mechanism for exchanging messages simultaneously. The two parties could be two businesspeople Alice and Bob negotiating a contract. They would each approve only if the other party approves. In case, if any party declines, the negotiation should fail. If Alice's and Bob's inputs are represented using two bits a and b respectively, the decision is equivalent to computing $a \wedge b$.

The privacy criteria require that Alice and Bob do not observe the input of the other party and can only observe the outcome $a \wedge b$ and whatever can be inferred from it. In the above example, this is important as Alice and Bob do not want to disclose their strategy to each other. The two parties may be physically remote and do not wish to engage a mediator who can inform the decision to them, because they do not wish to disclose their inputs to anyone.

We analyze the privacy criteria from Alice's point of view; analogous arguments would hold for Bob. If the decision $a \wedge b = 1$, then Alice knows with certainty that both her and Bob's inputs were 1. If Alice's input $a = 1$ and the decision $a \wedge b$ is 0, she knows that Bob's input was $b = 0$. This leaves Alice with the only possibility of interest, which is when her input is $a = 0$ and the decision will be $a \wedge b = 0$, irrespective of Bob's input b. In that case she should not learn if Bob's input b is equal to 0 or 1.

An OT protocol for this problem proceeds as follows. Alice has two binary inputs b_0 and b_1, and Bob has input in the form of a selection bit s. At the end of the protocol, Bob should obtain a single bit b_s, i.e., b_0 if $s = 0$ and b_1 if $s = 1$ from Alice. The privacy criteria require that Bob should not know Alice's other input b_{1-s} and Alice should not know Bob's input s. To solve the negotiation problem as discussed above, Alice sets $b_0 = 0$ and $b_1 = a$, and Bob sets $s = b$. It is easy to verify that at the end of the protocol, Bob will obtain $a \wedge b$ as output, which he can communicate to Alice after satisfying the corresponding privacy criteria. To satisfy fairness and correctness, we require that Bob accurately communicates the result to Alice, which implies that the parties are semi-honest.

Even et al. (1985) proposed the construction for the above problem using RSA primitives (Rivest et al. 1978b), which proceeds as follows.

1-2 Oblivious Transfer.
Inputs:

(a) Alice has binary inputs b_0 and b_1.
(b) Bob has binary input s.

Output: Bob obtains b_s.

1. Alice generates an RSA key pair that consists of two large primes p and q, and their product n, the public exponent e and the private exponent d. She sends n and e to Bob.
2. Alice generates two random numbers $x_0, x_1 \in \mathbb{Z}_n$ and sends them to Bob.
3. Bob chooses x_s. He generates a random number k and computes $v = (x_s + k^e)$ mod n, which he sends to Alice.
4. Alice subtracts x_0 and x_1 from v and applies the private exponent to obtain $k_0 = (v - x_0)^d \mod n$ and $k_1 = (v - x_1)^d \mod n$.
5. Alice blinds her inputs with these messages, $m_0 = b_0 + k_0$ and $m_1 = b_1 + k_1$, and sends these to Bob.
6. Bob selects $m_s - k$ as the output and communicates it to Alice.

In Step 4, Alice does not know whether Bob chose x_0 and x_1 to compute v, and therefore tries to unblind it using both x_0 and x_1. Only one of k_0 and k_1 will be equal to k, but its identity is again is unknown to Alice. She therefore blinds her input with both k_0 and k_1. Bob already knows which one out of m_0 and m_1 is blinded by k, and can unblind only that input. As Bob cannot blind the other input, the privacy of Alice's other input b_{1-s} is preserved, and similarly, as Alice does not know if Bob had selected x_0 or x_1, the privacy of Bob's input s is preserved.

1-2 OT can also be generalized to 1-n OT (Naor and Pinkas 1999; Aiello et al. 2001; Laur and Lipmaa 2007), where Alice has n inputs $\{b_0, \ldots, b_{n-1}$, and *Bob* has a selector integer $s = \{0, \ldots, n - 1\}$. Another extension is the k-n OT problem (Brassard et al. 1986), where Alice has n inputs, and Bob has k selectors $\{s_0, \ldots, s_{n-1}\}$.

Oblivious transfer is also related to the *private information retrieval* (PIR) problem (Chor et al. 1998). The main differences between the two problem is that in PIR, Alice does not care about the privacy of her inputs, and a non-trivial PIR protocol requires to have sub-linear bandwidth requirement. In terms of privacy, PIR can be considered a weaker version of OT. Ioannidis and Grama (2003) present a construction for the Millionaire problem (Yao 1982) and can be used as a protocol for the private comparison problem discussed at the beginning of Sect. 3.2.

Oblivious transfer is fundamental problem in cryptography. Kilian (1988), Crépeau et al. (1995) show that OT can be used to create a two-party SMC protocol for any function. These constructions are feasibility results and inefficient to be used in practice mainly due to high bandwidth costs. We, therefore, construct protocols using the techniques of homomorphic encryption and masking, and use oblivious transfer only when necessary.

3.2.8 Related Work on SMC Protocols for Machine Learning

There has been a lot of work on constructing privacy preserving algorithms for specific problems in privacy preserving data mining such as decision trees (Vaidya et al. 2008a), set matching (Freedman et al. 2004), clustering (Lin et al. 2005), association rule mining (Kantarcioglu and Clifton 2004), naive Bayes classification (Vaidya et al. 2008b), support vector machines (Vaidya et al. 2008c). Please refer to Vaidya et al. (2006), Aggarwal and Yu (2008) for a summary.

However, there has been somewhat limited work on SMC protocols for speech processing applications, which is one of the motivations for this thesis. Probabilistic inference with privacy constraints is a relatively unexplored area of research. The only detailed treatment of privacy-preserving probabilistic classification appears in Smaragdis and Shashanka (2007). In that work, inference via HMMs is performed on speech data using existing cryptographic primitives. The protocols are based on repeated invocations of privacy-preserving two-party maximization algorithms, in which both parties incur exactly the same protocol overhead.

3.3 Differential Privacy

The differential privacy model was originally introduced by Dwork (2006). Given any two databases D and D' differing by one element, which we will refer to as *adjacent databases*, a randomized query function M is said to be differentially private if the probability that M produces a response S on D is close to the probability that M produces the same response S on D' (Fig. 3.3). As the query output is almost the same in the presence or absence of an individual entry with high probability, nothing can be learned about any individual entry from the output. Differential privacy is formally defined as follows.

Definition 1. A randomized function M with a well-defined probability density P satisfies ε-differential privacy if, for all adjacent databases D and D' and for any set

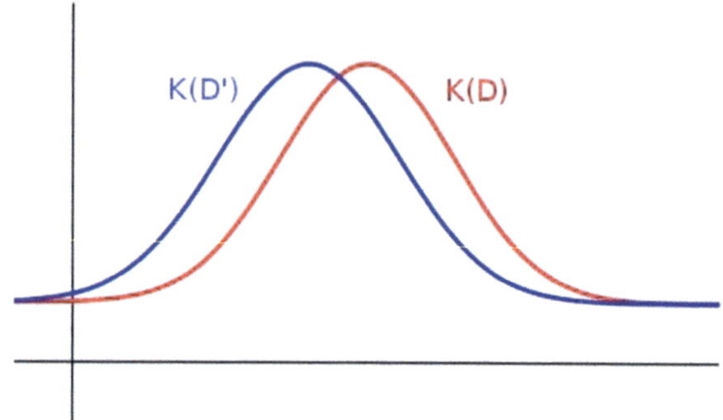

Fig. 3.3 Densities of mechanisms evaluated over adjacent datasets

of outcomes $S \in range(M)$,

$$\left| \log \frac{P[M(D) \in S]}{P[M(D') \in S]} \right| \leq \varepsilon. \tag{3.5}$$

Differential privacy provides the relative guarantee that the release of information will be just as likely whether or not the data about an individual is present in the database. As a consequence, if an individual chooses to contribute to the database, there is little or no increase in the privacy risk of the individual as opposed to not choosing to contribute to the database. For instance, if an insurance company is using a differentially private statistic computed over a database to decide whether or not to insure a particular person, the presence or absence of that person from the database will not significantly affect the chances of receiving coverage.

Differential privacy is an *ad omnia* guarantee as opposed to other models that provide *ad hoc* guarantees against a specific set of attacks and adversarial behaviors. This is a stronger guarantee than SMC where the only condition is that the parties are not able to learn anything about the individual data beyond what may be inferred from the final result of the computation. For instance, when the outcome of the computation is a classifier, it does not prevent an adversary from postulating about the presence of data instances whose absence might change the decision boundary of the classifier, and verifying the hypothesis using auxiliary information if any. Moreover, for all but the simplest computational problems, SMC protocols tend to be highly expensive, requiring iterated encryption and decryption and repeated communication of encrypted partial results between participating parties.

Differential privacy operates in two settings: *interactive* and *non-interactive*. In the interactive setting, the users repeatedly query the curator about information from the database, and the curator needs to modify the response in order to protect the privacy of the individual database entries. Differential privacy is composable: when consecutive N queries are executed, each satisfying ε-differential privacy, the combined set of queries satisfies $N\varepsilon$-differential privacy. ε can be thought of as a privacy cost incurred when answering an individual query. In the non-interactive setting, the curator computes a function on the dataset and publishes the response only once and the data is not used thereafter. In this thesis, we will almost exclusively look at non-interactive differentially private release mechanisms.

In a classification setting, the training dataset may be thought of as the database and the algorithm learning the classification rule as the query mechanism. A classifier satisfying differential privacy implies that no additional details about the individual training data instances can be obtained with certainty from the output of the learning algorithm, beyond the *a priori* background knowledge. An adversary who observes the values for all except one entry in the dataset and has prior information about the last entry cannot learn anything with high certainty about the value of the last entry beyond what was known *a priori* by observing the output. This can be thought to be an incentive to the dataset participants; nothing more about their data can be inferred from the differentially private classifier beyond what was already known about them.

3.3.1 Exponential Mechanism

Dwork et al. (2006) proposed the *exponential mechanism* for releasing continuous-valued functions satisfying ε-differential privacy. When a function f is evaluated over the dataset D, the true answer is $f(D)$. The mechanism M adds the appropriate kind of perturbation η depending on the sensitivity of f to $f(D)$ such that $f(D) + \eta$ satisfies differential privacy. The function sensitivity S is defined as follows.

Definition 2. For all adjacent databases D and D', the sensitivity S of the function f is given by

$$S = \max_{D,D'} \left\| f(D) - f(D') \right\|_1 . \tag{3.6}$$

The sensitivity S of a function f indicates how much is the function is likely to change after changing one instance from the dataset. For instance, the output of the function querying for maximum salary can change significantly after removing one instance: person having the maximum salary, and hence this function is highly sensitive. On the other hand, the output of the function counting the number of instances satisfying a particular property can change only by one and is relatively insensitive. It should be noted that the sensitivity is the property of the function, rather than any specific dataset.

Dwork et al. (2006) show that by adding perturbation η sampled from Laplace(S/ε) to the output of the function $f(D)$ satisfies ε-differential privacy. The variance of the perturbation η is proportional to S/ε; this means that we need to add perturbation with higher variance if our function is highly sensitive. Similarly, as the variance is inversely proportional to ε, we need to add perturbation with higher variance if our privacy constraint is tight. The perturbation introduces error as compared to the true answer which is inversely proportional to ε. This signifies the trade-off between privacy and accuracy.

3.3.2 Related Work on Differentially Private Machine Learning

The earlier work on differential privacy was related to functional approximations for simple data mining tasks and data release mechanisms (Dinur and Nissim 2003; Dwork and Nissim 2004; Blum et al. 2005; Barak et al. 2007). Although many of these works have connection to machine learning problems, more recently the design and analysis of machine learning algorithms satisfying differential privacy has been actively studied. Kasiviswanathan et al. (2008) present a framework for converting a general agnostic PAC learning algorithm to an algorithm that satisfies privacy constraints. Chaudhuri and Monteleoni (2008), Chaudhuri et al. (2011) use the exponential mechanism (Dwork et al. 2006) to create a differentially private logistic regression classifier by adding Laplace noise to the estimated parameters. They propose another differentially private formulation which involves modifying the objective function of the logistic regression classifier by adding a linear term scaled by Laplace noise. The second formulation is advantageous because it is independent of the classifier sensitivity which difficult to compute in general and it can be shown that using a perturbed objective function introduces a lower error as compared to the exponential mechanism.

However, the above mentioned differentially private classification algorithms only address the problem of binary classification. Although it is possible to extend binary classification algorithms to multi-class using techniques like one-vs-all, it is much more expensive to do so as compared to a naturally multi-class classification algorithm.

3.3.3 Differentially Private Speech Processing

In speech processing, differential privacy can be used as a publication mechanism for speech models such as Gaussian mixture models (GMMs) and hidden Markov models (HMMs). The advantage of publishing speech models is that it obviates the need for computationally expensive SMC protocols for training these models. Publishing differentially private classifiers, on the other hand, requires only a form of perturbation which introduces little or no computational overhead.

In practice, the problem with publishing differentially private speech models is that for meaningful privacy, e.g., $\varepsilon \leq 1$, the resulting classifiers provide very low accuracy on speech data. We present a mechanism for creating differentially private GMMs in Appendix A. The theoretical results show that the expected error is bounded by the square of the data dimensions. Even after parametrization, speech data is usually very high dimensional, typically with 39-dimensions and 100 frames per second: 3,900 features per training instance.

Speech processing applications such as speaker verification and speaker identification, however, require high accuracy as they are deployed in sensitive authentication and surveillance applications. The current state of the art methods of producing differentially private classifiers, such as the ones mentioned above, cannot be used in these applications. We outline ideas for creating usable differentially private classifiers in future work.

References

Aggarwal CC, Yu PS (eds) (2008) Privacy preserving data mining: models and algorithms. Advances in database systems, vol 34. Springer, New York

Aiello B, Ishai Y, Reingold O (2001) Priced oblivious transfer: How to sell digital goods. In: EUROCRYPT, pp 118–134

Barak B, Chaudhuri K, Dwork C, Kale S, McSherry F, Talwar F (2007) Privacy, accuracy, and consistency too: a holistic solution to contingency table release. In: Symposium on principles of database systems, pp 273–282

Barbaro M, Zeller T Jr. (2006) A face is exposed for AOL searcher no. 4417749. The New York Times, August 2006

Barth A, Datta A, Mitchell JC, Nissenbaum H (2006) Privacy and contextual integrity: framework and applications. In: IEEE Symposium on security and privacy, pp 184–198

Beerliová-Trubíniová Z, Hirt M (2008) Perfectly-secure MPC with linear communication complexity. In: Theory of cryptography, pp 213–230

Ben-Or M, Goldwasser S, Widgerson A (1988) Completeness theorems for non-cryptographic fault-tolerant distributed computation. In: Proceedings of the ACM symposium on the theory of computing, pp 1–10

Benaloh J (1994) Dense probabilistic encryption. In: First annual workshop on selected areas in cryptography, pp 120–128

Blum A, Dwork C, McSherry F, Nissim K (2005) Practical privacy: the suLQ framework. In: Symposium on principles of database systems

Blum M, Feldman P, Micali S (1988) Non-interactive zero-knowledge and its applications. In: ACM symposium on the theory of computing, pp 103–112

Boneh D, Goh E-J, Nissim K (2005) Evaluating 2-DNF formulas on ciphertexts. In: Kilian J (ed) Theory of cryptography conference 2005. LNCS, vol. 3378. Springer, Heidelberg, pp 325–341

Brassard G, Crépeau C, Robert J-M (1986) All-or-nothing disclosure of secrets. In: CRYPTO, pp 234–238

Chaudhuri K, Monteleoni C (2008) Privacy-preserving logistic regression. In: Neural information processing systems, pp 289–296

Chaudhuri K, Monteleoni C, Sarwate AD (2011) Differentially private empirical risk minimization. J Mach Learn Res 12:1069–1109

Chaum D, Crepeau C, Damgård I (1988) Multiparty unconditionally secure protocols. In: Simon J (ed) ACM symposium on the theory of computing, 1988. ACM Press, New York, pp 11–19

Chor B, Kushilevitz E, Goldreich O, Sudan M (1998) Private information retrieval. J ACM 45(6):965–981

Cramer R (1999) Introduction to secure computation. Lect Notes Comput Sci 1561:16–62

Cramer R, Damgård I, Nielsen JB (2001) Multiparty computation from threshold homomorphic encryption. In: EUROCRYPT, pp 280–300

Cramer R, Damgård I, Nielsen JB (2009) Multiparty computation, an introduction. http://cs.au.dk/jbn/smc.pdf

Crépeau C, van de Graaf J, Tapp A (1995) Committed oblivious transfer and private multiparty computation. In: Coppersmith D (ed) CRYPTO. LNCS, vol. 963, pp. 110–123. Springer, Heidelberg

Damgård I, Jurik M (2001) A generalisation, a simplification and some applications of paillier's probabilistic public-key system. In: Public key cryptography, pp 119–136

DeCew J (2008) Privacy. In: Zalta EN (ed) The stanford encyclopedia of philosophy. http://plato. stanford.edu/archives/fall2008/entries/privacy/

Diffie W, Hellman ME (1976) New directions in cryptography. IEEE Trans Inf Theory IT-22(6):644–654

Dingledine R, Mathewson N, Syverson PF (2004) Tor: the second-generation onion router. In: USENIX security symposium, USENIX, pp 303–320

Dinur I, Nissim K (2003) Revealing information while preserving privacy. In: ACM symposium on principles of database systems

Dwork C (2006) Differential privacy. In: International colloquium on automata, languages and programming

Dwork C, Nissim K (2004) Privacy-preserving datamining on vertically partitioned databases. In: CRYPTO

Dwork C, McSherry F, Nissim K, Smith A (2006) Calibrating noise to sensitivity in private data analysis. In: Theory of cryptography conference, pp 265–284

El-Gamal T (1985) A public-key cryptosystem and a signature scheme based on discrete logarithms. IEEE Trans Inf Theory 31:469–472

Even S, Goldreich O, Lempel A (1985) A randomized protocol for signing contracts. Commun ACM 28(6):637–647

Fontaine C, Galand F (2007) A survey of homomorphic encryption for nonspecialists. EURASIP J Inf security 2007:1–15

Freedman M, Nissim K, Pinkas B (2004) Efficient private matching and set intersection. In: Advances in cryptology—EUROCRYPT. Springer, Heidelberg, pp 1–19

Gavison R (1980) Privacy and the limits of law. Yale Law J 89:421–471

Gennaro R, Rabin MO, Rabin T (1998) Simplified VSS and fast-track multiparty computations with applications to threshold cryptography. In: ACM symposium on principles of distributed computing, pp 101–112

Gentry C (2009) Fully homomorphic encryption using ideal lattices. In: ACM symposium on theory of computing, pp 169–178

Gentry C (2010a) Toward basing fully homomorphic encryption on worst-case hardness. In: CRYPTO, pp 116–137

Gentry C (2010b) Computing arbitrary functions of encrypted data. Commun ACM 53(3):97–105

Goldreich O (2001) Foundations of cryptography: volume I basic tools. Cambridge University Press, New York

Goldreich O (2004) Foundations of cryptography: volume II basic applications. Cambridge University Press, New York

Goldwasser S, Micali S (1982) Probabilistic encryption and how to play mental poker keeping secret all partial information. In: ACM symposium on theory of computing, pp 365–377

Goldwasser S, Micali S (1984) Probabilistic encryption. J Comput Syst Sci 28(2):270–299

Goldwasser S, Micali S, Rackoff C (1985) The knowledge complexity of interactive proof-systems. In: ACM symposium on theory of computing, pp 291–304

Ioannidis I, Grama A (2003) An efficient protocol for Yao's millionaires' problem. In: Hawaii international conference on system sciences, pp 205–210

Kantarcioglu M, Clifton C (2004) Privacy-preserving distributed mining of association rules on horizontally partitioned data. IEEE Trans Knowl Data Eng 16(9):1026–1037

Kantarcioglu M, Kardes O (2008) Privacy-preserving data mining in the malicious model. J Inf Comput Secur 2(4):353–375

Kasiviswanathan SP, Lee HK, Nissim K, Raskhodnikova S, Smith A (2008) What can we learn privately? In: IEEE symposium on foundations of computer science, pp 531–540

Kilian J (1988) Founding cryptography on oblivious transfer. In: ACM symposium on theory of computing, pp 20–31

Koblitz N (1987) Elliptic curve cryptosystems. Math of Comput 48(177):203–209

Laur S, Helger L (2007) A new protocol for conditional disclosure of secrets and its applications. In: Applied cryptography and network security, pp 207–225

Lauter K, Naehrig M, Vaikuntanathan V (2011) Can homomorphic encryption be practical? In: ACM cloud computing security workshop

Lin X, Clifton C, Zhu MY (2005) Privacy-preserving clustering with distributed EM mixture modeling. Knowl Inf Syst 8(1):68–81

Menezes AJ, van Oorschot PC, Vanston SA (eds) (1996) Handbook of applied cryptography. CRC Press, Boca Raton

Miller VS (1985) Use of elliptic curves in cryptography. In: Advances in cryptology—CRYPTO '85, pp 417–426

Naor M, Pinkas B (1999) Oblivious transfer and polynomial evaluation. In: ACM symposium on theory of computing, pp 245–254

Narayanan A, Shmatikov V (2008) Robust de-anonymization of large sparse datasets. In: IEEE symposium on security and privacy, pp 111–125

Paillier P (1999) Public-key cryptosystems based on composite degree residuosity classes. In: EUROCRYPT

Quisquater J-J, Guillou LC, Berson TA (1989) How to explain zero-knowledge protocols to your children. In: CRYPTO, pp 628–631

Rabin M (1981) How to exchange secrets by oblivious transfer. Technical report TR-81, Harvard University

Rivest RL, Adleman L, Dertouzos ML (1978a) On data banks and privacy homomorphisms. In: Foundations of secure computation, pp 169–180

Rivest RL, Shamir A, Adleman L (1978b) A method for obtaining digital signatures and public-key cryptosystems. Commun ACM 21(2):120–126

Smaragdis P, Shashanka M (2007) A framework for secure speech recognition. IEEE Trans Audio Speech Lang Process 15(4):1404–1413

Vaidya J, Clifton CW, Yu Michael Z (2006) Privacy preserving data mining. Advances in information security, vol 19. Springer, New York

Vaidya J, Chris C, Kantarcioglu M, Patterson S (2008a) Privacy-preserving decision trees over vertically partitioned data. TKDD 2(3), 1–27

Vaidya J, Kantarcioglu M, Clifton C (2008b) Privacy-preserving naive bayes classification. VLDB J 17(4):879–898

Vaidya J, Hwanjo Y, Xiaoqian J (2008c) Privacy-preserving svm classification. Knowl Inf Syst 14(2):161–178

Warren SD, Brandeis LD (1890) The right to privacy. Harvard Law Rev 4:193–220

Westin AF (1967) Privacy and Freedom. Atheneum, New York

Yao A (1982) Protocols for secure computations (extended abstract). In: IEEE symposium on foundations of computer science

Part II
Privacy-Preserving Speaker Verification

Chapter 4
Overview of Speaker Verification with Privacy

4.1 Introduction

Speech being a unique characteristic of an individual is widely used as the biometric of choice in authentication systems. In an multi-factor authentication system, speech is usually used in combination with other items, something which the user has, e.g., a token or a smart card, and something which the user knows e.g., a personal identification number (PIN) or a password. Although the latter two are relatively easy to be compromised, speech, being a biometric, still remains robust to being compromised. We investigate text-independent speaker verification systems, that can authenticate while maintaining privacy over the speech data.

We consider a client-server model for speaker verification as shown in Fig. 4.1. The speaker verification system is the server, and the user executes a client program on a network-enabled computation device such as a computer or a smartphone. Speaker verification proceeds in two phases: enrollment, where the user submits a few speech utterances to the system, and verification where the user submits a test utterance and the system makes an acceptance decision based on its similarity to the enrollment data. Our primary privacy constraint is that the verification system should not be able to observe the speech samples provided by the user both during the enrollment and verification steps. This is important as the same speech data can potentially be used to verify the user in another authentication system. A speaker verification system might be compromised by an adversary who can access speech samples from the internal storage of the system for this purpose. Similarly, the system itself might be made to pose as a front to phish speech data from the unaware users. To prevent this, we require that the speech samples should be completely obfuscated from the system.

Our secondary privacy constraint is related to the computation device that is employed by the user in the verification process. We assume the user to be honest during the enrollment step, as it is in the user's privacy interest to be so. In the verification step, however, the user's computation device may be stolen or compromised by an adversary and can be used to impersonate the user. To prevent this, we require

M. A. Pathak, *Privacy-Preserving Machine Learning for Speech Processing*,
Springer Theses, DOI: 10.1007/978-1-4614-4639-2_4,
© Springer Science+Business Media New York 2013

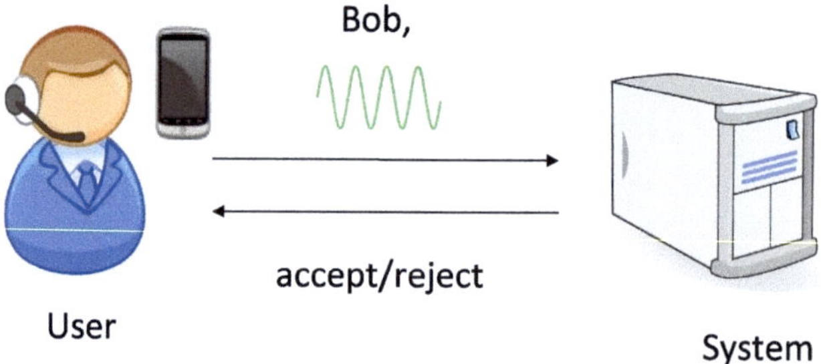

Fig. 4.1 Speaker verification work-flow

that no data that may allow the adversary to authenticate itself in place of the user should be stored on the device.

Our solution is based on two main ideas. Firstly, through the use of secure multiparty computation (SMC) protocols (Sect. 3.2) that enable the user and system to interact only on obfuscated speech data which the system cannot observe in plaintext at any point, we eliminate the possibility that the system could phish for a user's voice. Secondly, by requiring the system to store only obfuscated speaker models derived from speech data belonging to the user, we also ensure that an adversary who breaks into the system obtains no useful information. We assume that one adversary does not gain access to both the user's client device and the system at the same time.

We develop two frameworks for privacy-preserving speaker verification, satisfying the privacy constraints described above. In the first framework, which we describe in Chap. 5, we develop protocols for private enrollment and verification using Gaussian mixture models (GMMs) and homomorphic encryption. In the second framework which we which we describe in Chap. 6, we use the supervector representation with locality sensitive hashing (LSH) (Sect. 2.2.3) to transform the speaker verification problem into string comparison, and develop methods to perform it with privacy. The two frameworks provide a trade off between accuracy and efficiency: the GMM-based framework provides comparatively higher accuracy and the string comparison framework provides comparatively lower computational overhead. We investigate this trade off in detail in Chaps. 5 and 6.

We begin our discussion on privacy-preserving speaker verification, by examining the various privacy issues we mentioned above, that is relevant for both GMM and string comparison frameworks. We also consider the adversarial behavior of the various parties involved in the process.

4.2 Privacy Issues and Adversarial Behavior

We assume the user and the system to be independent parties that have access to separate computing devices operating in a client-server model and engaging in SMC protocols. We assume the parties to be computationally bounded, i.e., we consider that the parties cannot directly break the encryption or reverse the hash functions to obtain plaintext without knowing the decryption key or salt. In SMC, we consider two types of adversarial behavior of parties: *semi-honest* and *malicious*. We consider a semi-honest party to follow the steps of the protocol correctly, but keep a record of all intermediate results while trying to gain as much information as possible about the input data belonging to other parties. A malicious party, in addition to the semi-honest behavior, takes active steps in disrupting the protocol by using fraudulent input data to gain information about the input data belonging to other parties, and to obtain an unrealistic result.

Our main privacy constraint is that the system should not be able to observe the speech samples belonging to the user both during the enrollment and verification phases. The only malicious behavior the system can exhibit in the verification protocol is to modify the steps of the procedure to obtain incorrect output. As the system never observes the speech input in plaintext, this will not help it in anyway to obtain any information about the input. On the other hand, a system giving arbitrary accept/reject decisions will only antagonize the users and accepting false users would lead to a security problem, but not a loss in user privacy. We therefore assume that the system to be semi-honest. We also assume that the user is semi-honest during the enrollment phase. By maliciously submitting false user models, the user will only help in creating a weak authentication system, and there is no incentive for the user to do so. In the verification phase, however, we assume that the user is malicious, as the user could well be represented by an adversary who is trying to impersonate the user by using a compromised or stolen device. We make no assumptions about the correctness of the input data provided by the user; but we require that the system use the same input data for all models.

As discussed above, we consider two kinds of adversaries: having unauthorized access to the system data or the client device. The first type of adversary can use the system data to learn about the speech patterns of the user, violating user privacy, and also using it to impersonate the user in another speaker verification system. We therefore require the speech patterns and models to be encrypted or hashed by the user before being stored in the system. Similarly, the second type of adversary can use the data stored on the user's client device to impersonate the user. To prevent this, we require that no information about the speech patterns should be stored on the client device. However, we would need to store the user's cryptographic keys, on the client device, as these would later be required to encrypt the test input. If an adversary gains unauthorized access into both the client device and system data, we consider the system to be fully compromised as the adversary can use the cryptographic keys to decrypt the user models it obtains from the system.

4.2.1 Imposter Imitating a User

Apart from the adversarial behavior discussed above, an imposter can directly attempt to cheat a speaker verification system by imitating the voice of an enrolled user.

The imposter can also employ alternative mechanisms such as coercion, using a publicly available recording of a user's speech or morphing his/her own voice into the user's voice (Pellom and Hansen 1999; Sundermann et al. 2006). Although these type of attacks on speaker verification system do pose an important threat, we do not consider them as privacy violations, but as *security* issues. This is because the imposter does not gain any new information about the user's speech beyond what he/she had used to create the fake speech input. Therefore, the privacy of the user's speech is not compromised.

An imposter can also repeatedly interact with the system with different inputs, until he/she converges on an input that is similar to a legitimate user. This is less likely to succeed as the space of possible speech inputs is large. The system can block such users after a sufficient number of failures. Similarly, a verification system may arbitrarily deny a legitimate user, even though the user has submitted a speech sample. Again, for the same reasons we do not consider these as privacy violations.

4.2.2 Collusion

In SMC protocols, parties often collude with each other, by sharing their data, in order to gain information about the data belonging to other parties. In speaker verification, there are two types of parties: the system and the set of users. There is no motivation for the users to collude among themselves as the users only interact with the system in both enrollment and verification steps. By doing so they would not learn data about the speech input provided by other users. A user or a set of users could collude with the system, and the latter could transfer the enrollment data about the non-colluding users. This, however, would also not be useful, because the system only stores models in an obfuscated form: using encrypted speaker models in case of GMM-based protocols and cryptographically hashed keys in case of LSH-based protocols. In both the cases, the encryption and the hash function are applied by the user. In the absence of the respective private key or salt belonging to the genuine user, the colluding users will not be able to observe the original speech inputs. The system could also transfer the speech input submitted by the user in the verification phase. As the input is similarly encrypted or hashed by the user, this again will not help the system or the colluding user in obtaining information about the speech of the user.

4.2.3 Information Leakage After Multiple Interactions

A user typically enrolls with a speaker verification system only once, but performs the verification step multiple times. In each step, however, the user submits encrypted speech data in case of GMM-based protocols and cryptographically hashed keys in case of LSH-based protocols, both of which are obfuscated from the system.

As we shall see in Chap. 5, in GMM-based protocols, the user and the system execute a verification protocol and the system obtains two encrypted scores at the end of the verification protocols, corresponding to the speaker models and the UBM. In each execution, the system and the user can use different random inputs in the intermediate steps. After multiple interactions, the system would only observe a set of encrypted scores and this does not lead to any loss of user privacy.

As we shall see in Chap. 6, in LSH-based protocols, the system observes multiple cryptographically hashed LSH keys submitted by the user. Due to the one-way nature of the cryptographic hash function, the system is unable to correlate the hashes and gain information about the user input. The system only learns about the number of matched keys in each verification step. Again, this alone does not reveal any information about the user input.

References

Pellom BL, Hansen JHL (1999) An experimental study of speaker verification sensitivity to computer voice-altered imposters. In: IEEE international conference on acoustics, speech and signal processing

Sundermann D, Hoge H, Bonaforte A, Ney H, Black A, Narayanan S (2006) Text-independent voice conversion based on unit selection. In: IEEE international conference on acoustics, speech and signal processing

Chapter 5
Privacy-Preserving Speaker Verification Using Gaussian Mixture Models

In this chapter, we present an adapted UBM-GMM based privacy preserving speaker verification (PPSV) system, where the system is not able to observe the speech data provided by the user and the user does not observe the models trained by the system. We present protocols for speaker enrollment and verification which preserve privacy according to these requirements and report experiments with a prototype implementation on the YOHO dataset. We assume that the user represents the speech samples using mel-frequency cepstral coefficients (MFCC). We consider the MFCC features as the user input.

In the remainder of the chapter we describe the proposed protocols for private enrollment and verification in the PPSV system. We implement our protocols based on the commonly used UBM-GMM based speaker verification algorithm described in Bimbot et al. (2004) which we outline in Sect. 2.2.2. We describe the proposed protocols for privacy preserving speaker enrollment and verification in Sect. 5.2.

The operations on encrypted speech data motivated by our privacy requirements introduce significantly higher computation costs as compared to the non-private system. In Sect. 5.3, we report the increase in execution time and the effect on accuracy in experiments with a prototype PPSV system over the YOHO dataset (Campbell 1995).

5.1 System Architecture

Speaker verification proceeds in two separate phases: enrollment and verification. In the enrollment phase, each user submits enrollment samples to the system, and in the verification phase, a user submits a claimed identity along with a test sample which the system compares to the enrollment samples for the claimed user to arrive at the decision to accept/reject the user. The system uses the UBM and adapted model framework (Sect. 2.2.2) to represent the speaker model.

M. A. Pathak, *Privacy-Preserving Machine Learning for Speech Processing*,
Springer Theses, DOI: 10.1007/978-1-4614-4639-2_5,
© Springer Science+Business Media New York 2013

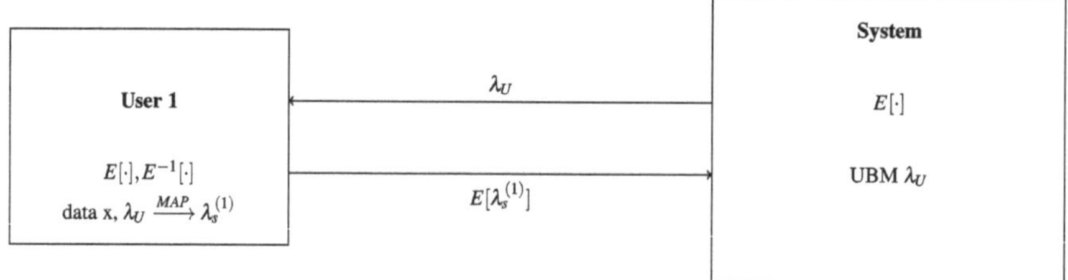

Fig. 5.1 Enrollment protocol: user has enrollment data x and system has the UBM λ_U. System obtains encrypted speaker model $E[\lambda_s^{(1)}]$

We present the design of the privacy preserving speaker verification system with a motivation to satisfy the privacy constraints discussed in Sect. 4.2. In the enrollment phase, the user and the system are assumed to be semi-honest. To start with, user generates a public/private key pair and sends the public key to the system. We assume that the system trains a UBM λ_U on publicly available data and stores it with itself as plaintext. In the enrollment protocol (Fig. 5.1), the system sends the UBM to the user in plaintext and the user performs the adaptation. The user then encrypts the adapted model with its key and sends it to the system. After executing the enrollment protocol with all users, the system has encrypted models for all users along with the UBM. At the end of the protocol, we require the user to delete the enrollment data from its computation device in order to protect it from an adversary who might gain unauthorized access to it. The user device only has the encryption and decryption keys. Similarly, as the server stores only the encrypted speaker models, it is also protected against an adversary who might compromise the system to gain the speaker models, in order to impersonate the user later. If an adversary compromises the user device as well as the system, we consider the system to be completely compromised as the adversary can use the decryption key to obtain the speaker model in plaintext.

In the verification protocol (Fig. 5.2), the user produces a test speech sample x and encrypts it using its key and sends it to the system along with the claimed identity. The system evaluates the encrypted test sample with the UBM and the encrypted model for the claimed speaker it had obtained in the enrollment protocol using the homomorphic operations and obtains two encrypted scores. The system makes its decision by comparing the difference between the two encrypted scores with a threshold using the compare protocol. This arrangement is sufficient for a semi-honest user, who provides the correct speech input while evaluating both the UBM and the speaker model. We construct the interactive private verification protocol in Sect. 5.2.2 to address semi-honest users.

In the verification phase, the user could also be malicious in the case it is represented by an imposter. A malicious user can gain an advantage in authenticating himself/herself by submitting different inputs during the evaluation of the UBM and

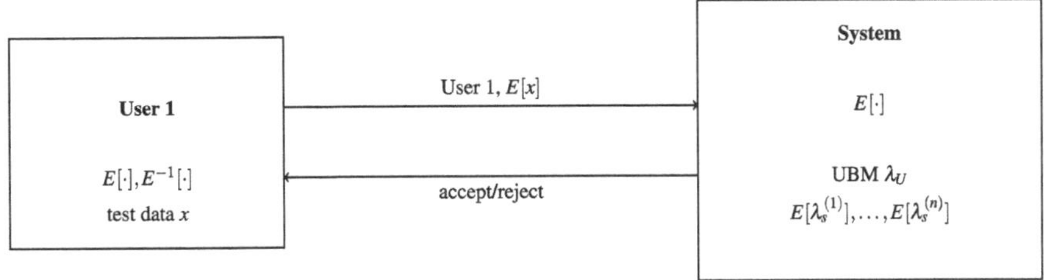

Fig. 5.2 Verification protocol: user has test data x and system has the UBM λ_U and encrypted speaker model $E[\lambda_s^{(1)}]$. The user submits encrypted data and the system outputs an accept/reject decision

the speaker models. We, therefore, need a protocol where the user gets to submit only one speech sample as input and the system can evaluate it on both the models without requiring any further participation from the user. We construct the non-interactive private verification protocol in Sect. 5.2.2 to address malicious users.

5.2 Speaker Verification Protocols

We now describe the enrollment and verification protocols in detail. We use the following construction from (Smaragdis and Shashanka, 2007): the multivariate Gaussian $\mathcal{N}(x; \mu, \Sigma)$ computed on any d-dimensional vector x can be represented in terms of a $(d + 1) \times (d + 1)$ matrix W.

$$\tilde{W} = \begin{bmatrix} -\frac{1}{2}\Sigma^{-1} & \Sigma^{-1}\mu \\ \hline 0 & w^* \end{bmatrix},$$

$$\text{where } w^* = -\frac{1}{2}\mu^T \Sigma^{-1}\mu - \frac{1}{2}\log|\Sigma|. \tag{5.1}$$

This implies $\log \mathcal{N}(x; \mu, \Sigma) = \tilde{x}^T \tilde{W}\tilde{x}$, where $\tilde{x}$ is an extended vector obtained by concatenating 1 to x. We reduce this computation to a single inner product $\bar{x}^T W$, where the extended feature vector $\bar{x}$ consists of all pairwise product terms $\tilde{x}_i \tilde{x}_j \in \tilde{x}$ and W is obtained by unrolling $\tilde{W}$ into a vector. In this representation

$$\log P(x|i) = \log \mathcal{N}(x; \mu, \Sigma) = x^T W \tag{5.2}$$

We assume that the user computes MFCC features from the speech samples. In the following discussion, we refer to the MFCC features as the speech sample itself.

5.2.1 Private Enrollment Protocol

We assume that the system already has access to the UBM, λ_U trained on a collection of publicly available speech data. The speaker verification algorithm requires a speaker model obtained from adapting the UBM to the enrollment data provided by the speaker. We require that the speaker model is kept with the system only after it is encrypted by the user's key. We outline this enrollment protocol below.

Private Enrollment Protocol.
Inputs:

(a) User has the enrollment samples $x_1, \ldots, x_n$ and both encryption key $E[\cdot]$ and decryption key $E^{-1}[\cdot]$.
(b) System has the UBM $\lambda_U = W_i^U$ for $i = 1, \ldots, N$, mixing weight α, and the encryption key $E[\cdot]$.

Output: system has the encrypted user model $E[\lambda_s] = E[\hat{W}_i^s]$, for $i = 1, \ldots, N$.

1. The system sends the UBM λ_U to the user.
2. User performs the model adaptation of λ_U with the enrollment samples $x_1, \ldots, x_n$ (see Sect. 2.2.2) to obtain the adapted model λ_s.
3. The user represents the mixture components of the adapted model using the $\hat{W}_i$ matrix representation described above.
4. The user encrypts $\hat{W}_i$ using its encryption key and sends it to the system.

Although this arrangement, where the user performs the adaptation, is adequate in most applications, we also develop the following protocol where the system performs the adaptation over encrypted enrollment data to obtain the encrypted speaker models (see Sect. 5.5).

5.2.2 Private Verification Protocols

In the verification protocol, the system needs to evaluate the probabilistic score of the given test sample using the UBM λ_U and the adapted model λ_s. This score is evaluated for all frames of the test sample; for a test sample $x = \{x_1, \ldots, x_T\}$ and the model λ, this score is given by

$$P(x|\lambda) = \prod_t \sum_j P(x_t|j) = \prod_t \sum_j w_j \mathcal{N}(x_t; \mu_j, \Sigma_j).$$

We compute this score in the log domain to prevent numerical underflow,

$$\log P(x|\lambda) = \sum_t \log \sum_j P(x_t|j) = \sum_t \log \sum_j e^{x_t^T W_j}, \tag{5.3}$$

using the W matrix representation from Eq. 5.2.

In our privacy model, we assume that the user has the speech sample x and the system has the encrypted matrices W_j. The verification protocol proceeds as follows: the user sends the encrypted frame vectors $E[x_t]$ to the system which the server uses to homomorphically compute the inner products $E[x_t^T W_j]$. In order to use the inner products to compute the log scores, we need to perform an exponentiation operation on ciphertext. As our cryptosystem only supports homomorphic additions and a single multiplication, it is not possible to do this directly, and we therefore use the *logsum protocol* which requires user participation in the intermediate steps. We outline this interactive verification protocol below.

Interactive Private Verification Protocol.
Inputs:

(a) User has the test sample x with frame vectors $\{x_1, \ldots, x_T\}$ and both encryption key $E[\cdot]$ and decryption key $E^{-1}[\cdot]$.
(b) System has $E[\lambda] = E[W_j]$, for $j = 1, \ldots, N$, and the encryption key $E[\cdot]$.

Output: System obtains the score $E[\log P(x|\lambda)]$.

1. The user encrypts the frame vectors $E[x_t]$ and sends it to the system.
2. For each mixture matrices $E[W_j]$ and each frame vector $E[x_t]$, the system computes the inner product $E[x_t^T W_j]$ using the *private inner product protocol*.
3. The system and the user then participate in the *logsum protocol* (Sect. 5.5) to obtain $E[\log P(x_t|\lambda)] = E[\log \sum_j e^{x_t^T W_j}]$.
4. The system adds the logsums homomorphically to obtain the $E[\log P(x|\lambda)]$.

As the system has access to the UBM in plaintext, the user and the system can execute the private mixture of Gaussians evaluation protocol (MOG) given by (Smaragdis and Shashanka, 2007). We however observe that the above protocol is substantially faster than MOG using unencrypted models. This is because in the above protocol, the user computes part of the inner products $x_t^T W_i^s$ in plaintext. We therefore repeat the above protocol with the encrypted UBM $E[W_i^U]$ to obtain the encrypted probability $E[\log P(x|\lambda_U)]$. The system and the user finally execute the *compare protocol*, to privately compute if $\log P(x|\lambda_U) > \log P(x|\lambda_s) + \theta$ and the system uses this as the decision to authenticate the user.

Throughout this protocol, the system never observes the frame vectors x_t in plaintext. The various supplementary protocols require the system to send encrypted partial results to the user. While doing so the system either adds or multiplies the values it sends to the user by a random number. This also prevents the user from learning anything about the partial results obtained by the system. Even after satisfying these privacy constraints, the system is able to make the decision on authenticating the user.

A disadvantage of the above protocol is that it requires participation from the user in the intermediate steps. Apart from the computational and data transfer overhead incurred, this also results in a privacy vulnerability. A malicious user can provide fake inputs in the logsum step to disrupt the protocol and try to authenticate itself without having a speech sample belonging to a genuine user. To prevent this behavior, we can augment the interactive protocols with zero-knowledge proofs (ZKP) and threshold decryption as discussed in Sect. 3.2.6. These constructions, however, have a significantly large overhead as compared to protocols using conventional Paillier encryption, such as the one above.

As feasibility is one of our aims, we instead construct a non-interactive protocol, where the user needs to submit the encrypted speech sample and the system directly computes the probability scores. As this is not possible due to the exponentiation involved in Eq. 5.3, we modify the score function itself by flipping the log and sum.

$$\text{score}(x, \lambda) = \sum_t \sum_j \log P(x_t|j) = \sum_t \sum_j x_t^T W_j. \tag{5.4}$$

This has the advantage that once the system homomorphically computes the inner products $E[x_t^T W_j]$, it only needs to homomorphic addition to compute the score without requiring any user participation. We experimentally observe that the accuracy of this score is close to that of the original probabilistic score. We outline the non-interactive verification protocol below.

Non-Interactive Private Verification Protocol.
Inputs:

(a) User has the test sample x with frame vectors $\{x_1, \ldots, x_T\}$ and both encryption key $E[\cdot]$ and decryption key $E^{-1}[\cdot]$.
(b) System has $E[\lambda] = E[W_j]$, for $j = 1, \ldots, N$, and the encryption key $E[\cdot]$.

Output: System obtains the score $E[\log P(x|\lambda)]$.

1. The user encrypts the frame vectors $E[x_t]$ and sends it to the system.
2. For each mixture matrix $E[W_j]$ and each frame vector $E[x_t]$, the system computes the inner product $E[x_t^T W_j]$ using the *private inner product protocol.*
3. The system adds the inner products homomorphically to obtain the $E[\text{score}(x, \lambda)]$.

The system executes the same protocol using the adapted model and the UBM using the same inputs it receives from the user in Step 1. Similar to the interactive protocol, the system executes the *compare protocol* with the user to decide whether to authenticate the user. The system never observes the frame vectors in plaintext in this protocol as well. As there is no user participation after the initial encrypted frame vectors are obtained, there is no loss of privacy of the user data.

5.3 Experiments

We present the results of experiments with the privacy preserving speaker verification protocols described above. We created prototype implementations of the interactive and non-interactive verification protocols in C++ using the pairing-based cryptography (PBC) library (pbc 2006) to implement the BGN cryptosystem and

(OpenSSL 2010) to implement the Paillier cryptosystem. We performed the experiments on a 2 GHz Intel Core 2 Duo machine with 3 GB RAM running 64-bit Ubuntu.

5.3.1 Precision

Both interactive and non-interactive protocols achieved the same final probability scores as the non-private verification algorithm up to 5 digits of precision.

5.3.2 Accuracy

We used the YOHO dataset (Campbell, 1995) to measure the accuracy of the two speaker verification protocols. We trained a UBM with 32 Gaussian mixture components on a random subset of the enrollment data and performed MAP adaptation with the enrollment data for individual speakers to obtain the speaker models. We evaluate the UBM and the speaker models on the verification data for the speaker as the true samples and the verification data for all other speakers as the imposter samples. We use equal error rate (EER) as the evaluation metric. An EER of $x\,\%$ implies that when the false accept rate is $x\,\%$, the false reject rate is also $x\,\%$. We observed an EER of 3.1 % for the interactive protocol and 3.8 % for the non-interactive protocol. This implies that there is only a marginal reduction in performance by modifying the scoring function.

5.3.3 Execution Time

We measured the execution times for the verification protocols using BGN encryption keys of sizes 256 and 512-bits.[1] In practice, 512-bit keys are used for strong security (Sang and Shen 2009). We use 256 and 1024 bit keys for Paillier cryptosystem.

We perform the verification of a 1 s speech sample containing 100 frame vectors using the UBM and the speaker models each containing 32 mixture components. The non-private verification algorithm required 13.79 s on the same input.

We use the Paillier cryptosystem for the interactive protocol and the BGN cryptosystem for the non-interactive protocol, as the private inner product is needed in the latter. We summarize the results in Tables 5.1 and 5.2 for the interactive and non-interactive protocols respectively. We observe that the interactive protocol is faster than the non-interactive protocol. This is due to the execution of the *private inner product* for each frame vector needed for the non-interactive protocol. The system

---

[1] For 512-bit keys, we choose the two prime numbers q_1 and q_2 each of 256-bits, such that their product $n = q_1 q_2$ is a 512-bit number.

Table 5.1 Execution time for the interactive protocol with Paillier cryptosystem

Steps	256-bit	1024-bit
Encrypting $x_1, \ldots, x_T$	136.64 s	7348.32 s
Evaluating speaker model	101.80 s	1899.39 s
Evaluating UBM	95.24 s	1189.98 s
Total+adapted	333.68 s	10437.69 s
	~5 min, 33 s	~2 h, 53 min

Table 5.2 Execution time for the non-interactive protocol with BGN cryptosystem

Steps	256-bit	512-bit
Encrypting $x_1, \ldots, x_T$	40.0 s	96.4 s
Evaluating speaker model	17450.3 s	77061.4 s
Evaluating UBM	19.5 s	77.8 s
Total	17509.8 s	77235.6 s
	~4 h, 52 min	~21 h, 27 min

requires to perform multiplicative homomorphic operations to obtain the inner product. These operations in turn require the computation of a bilinear pairing which is much slower than homomorphically multiplying plaintexts with ciphertexts as we do in the interactive protocol.

In both protocols, we observe that the UBM evaluation is significantly faster than the speaker model evaluation: this is because the UBM is available in plaintext with the system and the inner product requires only additive homomorphic operations. This is in contrast to evaluating the speaker model that is only available in ciphertext.

We observe that a vast amount of time is spent in homomorphically multiplying the encrypted frame vectors x_t and the encrypted model vectors W. The execution time of the protocol can be substantially reduced by using a faster implementation of the BGN cryptosystem, e.g., using a parallel computational framework such as GPUs. We leave this direction of experimentation for future work.

5.4 Conclusion

In this chapter, we developed the privacy-preserving protocol for GMM based algorithm for speaker verification using homomorphic cryptosystems such as BGN and Paillier encryption. The system observes only encrypted speech data and hence cannot learn anything about the user's speech. We constructed both interactive and non-interactive variants of the protocol. The interactive variant is relevant in the case of semi-honest adversary and the non-interactive variant is necessary in the case of malicious adversary.

During the exchanges required by the protocols, the user only observes additively or multiplicatively transformed data, and also cannot learn anything from it.

The proposed protocols are also found to give results which are identical, up to a high degree of precision as compared to a non-private GMM-based algorithm. The interactive protocol is more efficient than the non-interactive protocol as the latter requires homomorphic multiplication.

5.5 Supplementary Protocols

Private Enrollment Protocol with Encrypted Data.
Inputs:

(a) User has the MFCC feature vectors of the enrollment samples $x_1, \ldots, x_T$ and both encryption key $E[\cdot]$ and decryption key $E^{-1}[\cdot]$.
(b) System has the UBM $\lambda_U = W_i^U$ for $i = 1, \ldots, N$, mixing weight α, and the encryption key $E[\cdot]$.

Output: System has the encrypted adapted model $E[\lambda_s] = \{E[\hat{w}_i^s], E[\hat{\mu}_i^s], E[\hat{\Sigma}_i^s]\}$.
Computing the posterior probabilities
For each t:

1. The user computes the extended vector $\bar{x}_t$ from x_t and sends the encrypted vectors $E[\bar{x}_t]$ to the system.
2. The system computes the encrypted log probabilities for each mixture component
$$E\left[\log w_i^U P(x_t|i)\right] = \sum_j E[\bar{x}_{t,j}]^{W_{i,j}^U} + E[\log w_i^U].$$
3. The *logsum* protocol (Smaragdis and Shashanka, 2007) enables a party holding $E[\log x]$ and $E[\log y]$ to collaborate with another party who holds the private encryption key to obtain $E[\log(x + y)]$ without revealing x or y. The system participates with the user's client in the logsum protocol to obtain $E\left[\log \sum_i w_i^U P(x_t|i)\right]$.
4. The system computes encrypted log posteriors: $E[P(i|x_t)] = E\left[\log w_i^U P(x_t|i)\right] - E\left[\log \sum_i w_i^U P(x_t|i)\right]$.
5. The *private exponentiation protocol* enables a party holding $E[\log x]$ to collaborate with the party who holds the encryption key to obtain $E[x]$ without revealing x. The system then executes the private exponentiation protocol with the user to obtain $E[P(i|x_t)]$.

Learning w_i^s.

1. The system computes the encrypted value of w_i', $E[w_i']$ as $E\left[\sum_t P(i|x_t)\right] = \prod_t E[P(i|x_t)]$.
2. The system then computes the encrypted updated mixture weights as
$$E[\hat{w}_i^s] = E[w_i']^{\alpha/T} E[w_i^U]^{1-\alpha}.$$

Learning $\hat{\mu}_i^s$.

1. The system generates a random number r and homomorphically computes $E[P(i|x_t) - r]$. The system sends this quantity to the user.
2. The user decrypts this quantity and multiplies it by individual feature vectors x_t to obtain the vector $P(i|x_t)x_t - rx_t$. The user encrypts this and sends $E[P(i|x_t)x_t - rx_t]$ to the system along with encrypted feature vectors $E[x_t]$.
3. The system computes $E[P(i|x_t)x_t] = E[P(i|x_t)x_t] = E[x_t]^r + E[P(i|x_t)x_t - rx_t]$. It then computes $E\left[\sum_t P(i|x_t)x_t\right] = \prod_t E[P(i|x_t)x_t]$.
4. The *private division protocol* enables a party holding $E[x]$ and $E[y]$ to collaborate with the party holding the encryption key to obtain $E[x/y]$ without revealing x or y. The system and the user participate in a private division protocol with $E\left[\sum_t P(i|x_t)x_t\right]$ and $E[w']$ as inputs to obtain $E[\mu']$.
5. The system then computes the encrypted adapted mean as.

$$E[\hat{\mu}_i^s] = E[\mu_i']^\alpha E[\mu_i^U]^{1-\alpha}.$$

Learning $\hat{\Sigma}_i^s$.

This is similar to learning $\hat{\mu}_i^s$ and continues from Step 2 of that protocol.

1. The user multiplies $P(i|x_t) - r$ by the product of the feature vectors $x_t x_t^T$ to obtain $P(i|x_t)x_t x_t^T - rx_t x_t^T$.[2] The user encrypts this and sends $E[P(i|x_t)x_t x_t^T - rx_t x_t^T]$ to the system along with encrypted feature vectors $E[x_t x_t^T]$.
2. The system computes $E[rx_t x_t^T] = E[x_t x_t^T]^r$ and multiplies it to $E[P(i|x_t)x_t x_t^T - rx_t x_t^T]$ to obtain $E[P(i|x_t)x_t x_t^T]$. The system multiplies these terms for all values of t to obtain $E\left[\sum_t P(i|x_t)x_t x_t^T\right]$.
3. The system engages the user participate in the private division protocol with this encrypted sum and $E[w']$ as inputs to obtain $E[\Sigma_i']$.
4. The system and the user participate in a *private vector product protocol* with input $E[\hat{\mu}_i^s]$ to obtain $E[\hat{\mu}_i^s \hat{\mu}_i^{sT}]$. The system then computes the encrypted updated covariance as

$$E[\hat{\Sigma}_i^s] = E[\Sigma_i']^\alpha E[\Sigma_i^U + \mu_i^U \mu_i^{UT}]^{1-\alpha} - E[\hat{\mu}_i^s \hat{\mu}_i^{sT}].$$

The encrypted model parameters obtained by the system cannot be directly used in the verification protocol. We can then compute the encrypted model matrix $E[\hat{W}_i^s]$ from the encrypted model parameters. At no point in the enrollment protocol, the system observes feature vectors x_t in plaintext and the adapted models obtained by the system are also encrypted.

Model Construction Protocol.

Inputs:

(a) User has encryption key $E[\cdot]$ and decryption key $E^{-1}[\cdot]$

---

[2] For efficiency purposes, only the diagonal terms of the product $x_t x_t^T$ can be included without having a significant impact on accuracy.

(b) System has encrypted adapted model parameters $E[\lambda_s] = \{E[\hat{w}_i], E[\hat{\mu}_i], E[\hat{\Sigma}_i]\}$, for $i = 1, \ldots, N$, and the encryption key $E[\cdot]$.

Output: System has the encrypted adapted model matrices $E[\hat{W}_i]$, for $i = 1, \ldots, N$.

1. For each covariance matrix $\hat{\Sigma}_i$, the system generates a random number q and homomorphically multiplies it with $E[\hat{\Sigma}_i]$ and sends the result $E[q\hat{\Sigma}_i]$ to the user.
2. The user decrypts this quantity and obtains the matrix $q\hat{\Sigma}_i$. The user then computes the reciprocal matrix $\frac{1}{q}\hat{\Sigma}_i^{-1}$. The user encrypts this matrix and sends $E[\frac{1}{q}\hat{\Sigma}_i^{-1}]$ to the system.
3. The system homomorphically multiplies the encrypted matrix by $-\frac{q}{2}$ to get the encrypted matrix $E[-\frac{1}{2}\hat{\Sigma}_i^{-1}]$.
4. The user also computes the log determinant of the matrix $\frac{1}{q}\hat{\Sigma}_i^{-1}$ to obtain $-\frac{1}{q}\log|\hat{\Sigma}_i|$, which the user encrypts and sends to the system.
5. The system homomorphically multiplies the encrypted log determinant by $\frac{q}{2}$ to obtain $E[-\frac{1}{2}\log|\hat{\Sigma}_i|]$.
6. The system generates a random $d \times 1$ matrix r and homomorphically computes $E[\hat{\mu}_i - r]$ and sends it to the user.
7. The user decrypts this vector to obtain $\hat{\mu}_i - r$ and multiplies it with $\frac{1}{q}\hat{\Sigma}_i^{-1}$ computed in Step 3 to obtain $\frac{1}{q}\hat{\Sigma}_i^{-1}\hat{\mu}_i - \frac{1}{q}\hat{\Sigma}_i^{-1}r$. The user encrypts this and sends it to the system.
8. Using the matrix r and $E[\frac{1}{q}\hat{\Sigma}_i^{-1}]$ which was obtained in Step 4, the system homomorphically computes $E[\frac{1}{q}\hat{\Sigma}_i^{-1}r]$ and adds it homomorphically to the encrypted vector obtained from the user to get $E[\frac{1}{q}\hat{\Sigma}_i^{-1}\hat{\mu}_i]$. The system then homomorphically multiplies this by q to get $E[\hat{\Sigma}_i^{-1}\hat{\mu}_i]$.
9. The system executes the encrypted product protocol using $E[\hat{\Sigma}_i^{-1}\hat{\mu}_i]$ and $E[\hat{\mu}_i]$ as inputs to obtain $E[\hat{\mu}_i^T\hat{\Sigma}_i^{-1}\hat{\mu}_i]$ and multiplies it homomorphically with $-\frac{1}{2}$ to obtain $E[-\frac{1}{2}\hat{\mu}_i^T\hat{\Sigma}_i^{-1}\hat{\mu}_i]$.
10. Finally, the system homomorphically adds the encrypted scalars obtained above along with $E[\hat{w}_i]$ to get $E[w^*]$.
    The system thus obtains all the components necessary to construct $E[\hat{W}_i]$.

Logsum Protocol.
Inputs:

(a) Alice has both encryption key $E[\cdot]$ and decryption key $E^{-1}[\cdot]$.
(b) Bob has $E\left[\log z_i\right]$ for $i = 1, \ldots, N$, and the encryption key $E[\cdot]$.

Output: Bob has $E\left[\log \sum_i z_i\right]$

1. Bob generates a random number r and homomorphically computes

$$E[\log z_i - r] = E[\log z_i] - E[r].$$

He sends this to Alice.

2. Alice decrypts this quantity and exponentiates it to obtain $z_i e^{-r}$.
3. Alice adds these quantities to compute $e^{-r} \sum_i z_i$ and then computes the log to obtain $\log \sum_i z_i - r$.
4. Alice encrypts $E\left[\log \sum_i z_i - r\right]$ and sends it to Bob.
5. Bob homomorphically adds r to obtain the desired output.

$$E\left[\log \sum_i z_i\right] = E\left[\log \sum_i z_i - r\right] + E[r].$$

Compare protocol.
Inputs:

(a) Alice has the encryption keys $E_A[\cdot]$, $E_B[\cdot]$ and decryption key $E_A^{-1}[\cdot]$.
(b) Bob has $E_A[x]$ and $E_A[y]$ and the encryption keys $E_A[\cdot]$, $E_B[\cdot]$ and decryption key $E_B^{-1}[\cdot]$.

Output: Bob knows if $x > y$.

1. Bob generates a random positive or negative number q and multiplicatively masks $E_A[x - y]$ to obtain $E_A[q(x - y)]$. He sends this quantity to Alice.
2. Alice decrypts this quantity to obtain $q(x - y)$. She generates a positive random number r and computes $qr(x - y)$ and encrypts this using Bob's key to obtain $E_B[qr(x - y)]$.
3. Bob decrypts this quantity and divides it by q. Bob checks for $r(x - y) > 0$ and uses that to conclude whether $x > y$.

References

Bimbot F, Bonastre J-F, Fredouille C, Gravier G, Magrin-Chagnolleau I, Meignier S, Merlin T, Ortega-Garcia J, Petrovska-Delacretaz D, Reynolds D (2004) A tutorial on text-independent speaker verification. EURASIP J Appl Signal Process 4:430–451

Campbell JP (1995) Testing with the YOHO CD-ROM voice verification corpus. In: IEEE international conference on acoustics, speech and signal processing, pp 341–344

OpenSSL (2010) http://www.openssl.org/docs/crypto/bn.html

pbc. PBC library (2006) http://crypto.stanford.edu/pbc/

Sang Y, Shen H (2009) Efficient and secure protocols for privacy-preserving set operations. ACM Trans Inf Syst Secur 13(1):9:1–9:35

Smaragdis P, Shashanka M (2007) A framework for secure speech recognition. IEEE Trans Audio Speech Lang Process 15(4):1404–1413

Chapter 6
Privacy-Preserving Speaker Verification as String Comparison

In this chapter, we develop a method for speaker verification that requires minimal computation overhead needed to satisfy the privacy constraints. The central aspect of our approach is to reduce the speaker verification task to string comparison. Instead of using the UBM-GMM approach, we convert the utterances into supervector features (Campbell et al. 2006) that are invariant with the length of the utterance. By applying the locality sensitive hashing (LSH) transformation (Gionis et al. 1999) to the supervectors, we reduce the problem of nearest-neighbor classification into string comparison. It is very efficient to perform string comparison with privacy, similar to a conventional password system. By applying a cryptographic hash function, e.g., SHA-256 (SHA 2008), we convert the LSH transformation to an obfuscated string which the server cannot use to gain information about the supervectors, but is still able to compare if two strings are identical. This one-way transformation preserves the privacy of the speech utterances submitted by the user, and can be executed significantly faster than applying homomorphic encryption.

We emphasize that our main goal is not to develop speaker verification algorithms that achieve higher accuracy. We are principally interested in developing an efficient speaker verification system that satisfies the same privacy constraints but with a minimal computational overhead, while achieving feasible accuracy. Using the LSH functions defined over Euclidean and cosine distances, we show that our system achieves an equal error rate (EER) of 11.86 % on the YOHO dataset, while requiring only a few milliseconds of computational overhead. We, nevertheless, consider this result as a proof of concept; we believe the EER can be further improved by using more discriminative speech features, better quality training data, and supervectors computed over a larger number of Gaussian components, all while utilizing the same private string comparison framework. Although UBM-GMM and SVMs trained over supervectors are known to supersede our current accuracy, our proposed algorithm significantly supersedes the computational overhead of the privacy-preserving variants of these algorithms (Fig. 6.1).

M. A. Pathak, *Privacy-Preserving Machine Learning for Speech Processing*, 67
Springer Theses, DOI: 10.1007/978-1-4614-4639-2_6,
© Springer Science+Business Media New York 2013

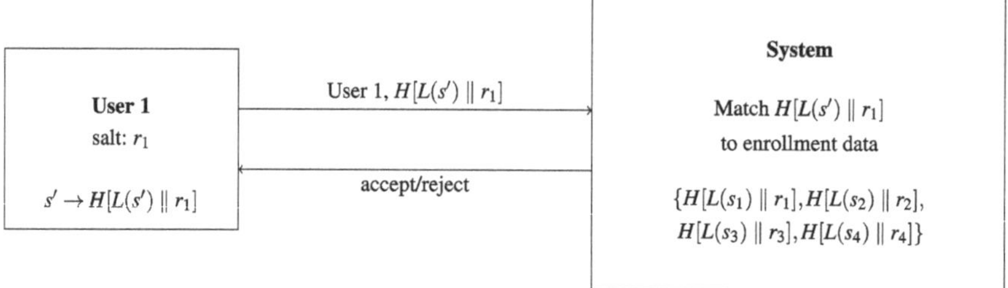

Fig. 6.1 System architecture. For user 1, test utterance supervector: s', salt: r_1. Although only one instance of LSH function L is shown, in practice we use l different instances

6.1 System Architecture

Speaker verification proceeds in two distinct phases: enrollment, where each user submits the speech data to the system, and verification, where a user submits a test utterance to the system which computes a yes/no decision as output. As discussed in Sect. 2.2.3, the users convert their speech utterances into supervectors and then apply l LSH functions. To facilitate this, the system provides a UBM trained on publicly available data along with the random feature vectors constituting the LSH functions. There is no loss of privacy in publishing these elements, as none of these depend on the speech data belonging to the user.

Our first privacy constraint is that the system should not be able to observe both the enrollment data and the test input provided by the user. As we discussed in Sect. 2.2.4, LSH is not privacy preserving. Due to the locality sensitive property of LSH, it is possible to reconstruct the input vector by observing a sufficient number of LSH keys obtained from the same vector.

We satisfy our privacy constraint by requiring the users to apply a cryptographic hash function $H[\cdot]$ to the LSH computed over the supervector s, which we denote by $H[L(s)]$. Cryptographic hash functions satisfy the property that they can be computed efficiently, i.e., in polynomial time on all inputs, but are computationally hard to invert, i.e., there is no feasible algorithm to obtain the input $L(s)$ for a given output $H[L(s)]$. Cryptographic hash functions, such as SHA-256, MD5, are orders of magnitude faster to compute as compared to homomorphic encryption, such as the Paillier cryptosystem.

Salted Cryptographic Hash

As the possible values for LSH keys lie in a relatively small set by cryptographic standards, 256^k for k-bit Euclidean LSH and 2^k for k-bit cosine LSH, it is possible for the server to obtain $L(s)$ from $H[L(s)]$ by applying brute-force search. To make this attack infeasible, we vastly increase the domain of the hash function $H[\cdot]$ by concatenating the LSH key with a long random string r_i (e.g., 80-bits in length) that is unique to the user i, which we refer to as the *salt*. By requiring the user to keep the salt private and unique to each system, this also gives us the additional advantage of the cryptographically hashed enrollment data being rendered useless to an adversary. An adversary who has gained access to the system data will be unable to use the hashed enrollment data while trying to impersonate the user in another system.

Our secondary privacy constraint is that an adversary should not be able to impersonate the user by having access to the compromised or stolen computation device belonging to that user. To satisfy this, we require the device to have no record of the enrollment samples and the previously used test samples. Although the salt r_i is stored on the device, it does not result in any loss of privacy by itself. The only point of failure in the system is when both the user device and the server data is compromised by the same adversary, who can then use the salt and the cryptographically hashed enrollment data to obtain the original LSH keys via brute-force search.

In the verification phase, the user could behave maliciously, as in reality, it could be an imposter trying to impersonate the user. The private verification protocols using LSH and cryptographic hash functions are non-interactive; we do not require the user to submit anything apart from the hashes computed over the test input. This automatically addresses malicious users. If an imposter submits malformed input in order to cheat the system into authentication, the hashes computed over that input are not likely to match the hashes previously submitted by the user in the enrollment step. This is due to the property of the LSH keys; the test input needs to be similar to the enrollment samples for the matches to take place. As discussed in Sect. 4.2.1 we do not consider an adversary using publicly data to create a reasonable imitation of a user's speech as a privacy violation.

6.2 Protocols

We present the details of the enrollment and verification protocols below.

A. **Enrollment Protocol**

Each user is assumed to have a set of enrollment utterances $\{x_1, \ldots, x_n\}$. The users also obtain the UBM and the l LSH functions $\{L_1(\cdot), \ldots, L_l(\cdot)\}$, each of length k-bit from the system. Each user i generates the random 80-bit salt string r_i.

For each enrollment utterance x_j, user i:

   a. performs adaptation of x_j with the UBM to obtain supervector s_j.
   b. applies the l LSH functions to s_j to obtain the keys $\{L_1(s_j), \ldots, L_l(s_j)\}$.
   c. applies the cryptographic hash function salted with r_i to each of these keys to obtain
      $\{H[L_1(s_j) \parallel r_1], \ldots, H[L_l(s_j) \parallel r_1]\}$, and sends them to the server.

B. **Verification Protocol**

For a test utterance x', user i:

    i. performs adaptation of x' with the UBM to obtain supervector s'.

   ii. applies the l LSH functions to s' to obtain the keys
$\{L_1(s'), \ldots, L_l(s')\}$.

  iii. applies the cryptographic hash function salted with r_i to each of these keys
to obtain $\{H[L_1(s') \parallel r_1], \ldots, H[L_l(s') \parallel r_1]\}$, and sends it to the server.

  iv. The server compares the hashed keys for the test utterance with the hashed
keys of the enrollment utterances, and counts the number of matches.
Depending on whether this number is above or below a threshold, the server
makes an accept/reject decision.

As discussed in Sect. 2.2.3, we consider two vectors to be matched if any one of
their LSH key matches. The server fine-tunes the acceptance threshold by experimenting with match counts on held-out data.

While the above enrollment and verification protocols only consider enrollment
data belonging to the speaker, it is also possible to include imposter data in the
enrollment step. The user can similarly obtain supervectors from publicly available
imposter data, apply LSH and the salted cryptographic hash function, and submit
the hashed keys to the server. In the verification protocol, the server can match the
test input separately to the hashed keys belonging to both the user enrollment and
imposter sets, and make the decision by comparing the two scores. In Sect. 6.3, we
observe that there is an improvement in performance by including the imposter data.

The server never observes any LSH key before a salted cryptographic hash function is applied to it. Apart from the salt, the user does not need to store any speech
data on its device. The enrollment and verification protocols, therefore, satisfy the
privacy constraints discussed above.

6.3 Experiments

We experimentally evaluate the privacy preserving speaker verification protocols
described above for accuracy and execution time. We perform the experiments on
the YOHO dataset (Campbell 1995).

6.3.1 Accuracy

We used Mel-frequency cepstral coefficient (MFCC) features augmented by differences and double differences, i.e., a recording x consists of a sequence of
39-dimensional feature vectors $x_1, \ldots, x_T$. Although in practice the UBM is supposed to be trained on publicly available data, for simulation, we trained a UBM
with 64 Gaussian mixture components on a random subset of the enrollment data

Table 6.1 Average EER for the two enrollment data configurations and three LSH strategies: Euclidean, cosine, and combined (Euclidean and cosine)

Enrollment: only speaker

Euclidean	Cosine	Combined
15.18%	17.35%	13.80%

Enrollment: speaker and imposter

Euclidean	Cosine	Combined
15.16%	18.79%	11.86%

belonging to all users. We obtained the supervectors by individually adapting all enrollment and verification utterances to the UBM.

We use *equal error rate* (EER) as the evaluation metric. We observed that the lowest EER was achieved by using $l = 200$ instances of LSH functions each of length $k = 20$ for both Euclidean and cosine distances. A test utterance considered to match an enrollment utterance if at least one of their LSH keys matches. The score for a test utterance is given by the number of enrollment utterances it matched. We report the average EER for different speakers in Table 6.1. We observe that LSH for Euclidean distance performs better than LSH for cosine distance. We also used combined LSH scores for Euclidean and cosine distances and found that this strategy performed the best. This can be attributed to the fact that different distance measures find approximate nearest neighbors in different parts of the feature space. We hypothesize that the EER can be further improved by combining LSH functions defined over an ensemble of diverse distance metrics. We leave a study in this direction for future work.

We also consider two configurations where only the enrollment data of the speaker was used, or secondly, where imposter data chosen randomly from the enrollment data of other speakers was used along with the enrollment data of the speaker. In the latter experiment, we compare the difference between the number of utterances matched by the test utterance in the enrollment and the imposter set. We observed that using imposter data achieved lower EER when using the combined scores for both the distances.

6.3.2 Execution Time

As compared to a non-private variant of a speaker recognition system based on supervectors, the only computational overhead is in applying the LSH and salted cryptographic hash function. For a $64 \times 39 = 2496$-dimensional supervector representing a single utterance, the computation for both Euclidean and cosine LSH involves a multiplication with a random matrix of size 20×2496 which requires a fraction of a millisecond. Performing this operation 200 times required 15.8 ms on average. We performed all the execution time experiments on a laptop running 64-bit Ubuntu 11.04 with 2 GHz Intel Core 2 Duo processor and 3 GB RAM.

The Euclidean and cosine LSH keys of length $k = 20$ require 8×20 bits = 20 bytes and 20 bits = 1.6 bytes for storage respectively. Using our C++ implementation of SHA-256 cryptographic hashing algorithm based on the OpenSSL libraries (OpenSSL 2010), hashing 200 instances of each of these keys in total required 28.34 ms on average. Beyond this, the verification protocol only consists of matching the 256-bit long cryptographically hashed keys derived from the test utterance to those obtained from the enrollment data.

6.4 Conclusion

In this chapter, we presented a framework for speaker verification using supervectors and LSH. The enrollment and verification protocols add a very small overhead of a few milliseconds to the non-private computation. This overhead is significantly smaller than the overhead of secure multiparty computation approaches using homomorphic encryption such as the GMM verification system, while satisfying the same privacy constraints. Experimentally, we observed that the algorithm achieves an equal error rate (EER) of 11.86 % in the case of using imposter data along with the speaker data and 13.80 % when only using speaker data. These EERs are higher than that of using UBM-GMM for speaker verification, but they can potentially be improved by engineering better supervector features. In general, the choice between using LSH and GMM represents a trade-off between speed and accuracy.

References

Campbell JP (1995) Testing with the YOHO CD-ROM voice verification corpus. In: IEEE international conference on acoustics, speech and signal processing

Campbell WM, Sturim DE, Reynolds DA, Solomonoff A (2006) SVM based speaker verification using a GMM supervector kernel and NAP variability compensation. In: IEEE international conference on acoustics, speech and signal processing

Aristides G, Piotr I, and Rajeev M (1999) Similarity search in high dimensions via hashing. In: Proceedings of the twenty-fifth international conference on very large databases, pp 518–529

SHA (2008) FIPS 180–3: Secure hash standard. National Institute for Standards and Technology

OpenSSL (2010). http://www.openssl.org/docs/crypto/bn.html

Part III
Privacy-Preserving Speaker Identification

Chapter 7
Overview of Speaker Identification with Privacy

7.1 Introduction

Speaker identification is the task of determining the identity of the speaker given a test speech sample. We typically associate the identity of the speaker with a predetermined set of speakers and we refer to this task as closed-set speaker identification. If we augment the set of speakers with a none of the above option, i.e., consider the speaker to be outside the predetermined set of speakers, this task becomes open-set speaker identification. Speaker verification can be considered as a generalization of open-set speaker identification, where the set of speakers is restricted to one speaker. As a consequence, the algorithms used in speaker verification can be extended for speaker identification. The main difference between the two tasks is their application scenarios and this results in different evaluation metrics for the two tasks. Speaker verification is mainly used for authentication; speaker identification finds use in surveillance applications and also as a preliminary step for other speech processing applications as we discuss below.

7.1.1 Speech-Based Surveillance

A recorded sample of conversational speech from forms a much stronger piece of evidence as compared to just a transcript. Because of this, security agencies, such as the police perform audio-based surveillance to gather information from speech samples spoken by the suspects. We refer to the parties performing the surveillance as the agency. The audio-based surveillance can be in the form of wiretapping, where the agency listens in on telephone conversations. The agencies also perform physical surveillance, by placing hidden microphones in public areas in order to eavesdrop on individuals. A basic characteristic of surveillance is that the agency needs to perform it obliviously, i.e., the subjects under surveillance should not know about

M. A. Pathak, *Privacy-Preserving Machine Learning for Speech Processing*, Springer Theses, DOI: 10.1007/978-1-4614-4639-2_7, © Springer Science+Business Media New York 2013

it. Although listening in on personal conversations either over the telephone or from physical sources is an effective surveillance method to identify credible security threats, this directly infringes on the privacy of innocent individuals, who may not be intended targets of the surveillance. To prevent this, the agency first needs to check if the given speech recording belongs to a speaker who is subject to surveillance. In order to do so, the agency would need to first listen to the conversations, which itself would cause the privacy violation, hence the circular problem.

One strategy to avoid such privacy violations is for the agency to perform privacy-preserving speaker identification on the surveillance recordings. The primary requirement of such privacy-preserving algorithms is that the agency is able to identify speaker corresponding to the speech sample without observing the speech sample. If the speaker is under surveillance, the agency can then demand the original speech sample from the party having the speech recording, e.g., the telephone company. In a criminal investigation setting, there is a legal authorization for this demand if the agency has a warrant for wiretapping against the identified speaker. An important consideration here is that if there is no warrant against the identified speaker, the agency cannot demand the phone recording. This protects the privacy of the telephone subscribers, as the telephone company can provide an assurance that their phone conversations will only be released only if there is a wiretapping warrant for the subscriber. This forms an effective balance between individual privacy of the subscribers and the audio-based surveillance necessary for security enforcement.

7.1.2 Preliminary Step for Other Speech Processing Tasks

Speaker identification is also used as the first step in more complex speech processing tasks, e.g., speech recognition, where we use separate recognition models trained on data for individual speakers. Given a speech sample, we first identify the speaker corresponding to it and then apply the recognition model for that speaker. This arrangement is known to provide higher accuracy as compared to applying a recognition model for a generic speaker.

In many cases we perform speech processing in a distributed client-server setting, where the client user has access to the speech sample and the server has access to the speaker models, for both identification and recognition. Due to the privacy and concerns, many users typically do not wish to export their speech data to an external party and this reduces the utility of the client-server model. By using privacy-preserving speaker identification and speech recognition protocols, the server can perform the necessary operations without observing the speech sample, thereby preserving the privacy constraints.

In the following two chapters, we develop protocols for privacy-preserving speaker identification using GMM and supervector based algorithms. We begin the discussion by formally considering the privacy issues in speaker identification in the context of a multiparty model below.

7.2 Privacy Issues and Adversarial Behavior

We consider speaker identification with two parties: the client who has access to the test speech sample and the server who has access to the speaker models and is interested in identifying the most likely speaker corresponding to the test sample. In case of wiretapping, the server would be the security agency, and the client would be the telephone company that has access to the conversations of all subscribers. It is important to note that we do not consider the individual phone users as the client. This is because they are supposed to be unaware about being subjected to surveillance, as no user would willingly participate in such an activity.

The agency has access to the speaker models for the individuals it already has wiretapping warrants against. The agency directly deals with the telephone company in order to identify if any such individual is participating in a telephone conversation. If that is found to be the case, the agency would follow the necessary legal procedure to obtain the phone recording. The same analogy also holds for the case of physical surveillance; e.g., a supermarket installs hidden security cameras with microphones in their premises, and the police might want the audio recordings to gain evidence about criminal activity. In this case the supermarket would be the client and the police would be the server. Although we aim to construct mechanisms to protect the privacy of the client data, we assume that the client cooperates with the server. In the speaker identification task, it is possible that the client can refuse the server by simply not providing the speech input or providing white noise or speech from some other source as input. In this case, it will not be possible for the server to perform surveillance. In order to prevent this, the server can legally require the client to use only the correct data as input. We also assume that the server already knows the K speakers it is looking to identify. The server uses publicly available data for the K speakers to train the corresponding speaker models without requiring the participation of the client. The server trains a universal background model (UBM) (Sect. 2.2.2) to handle the none of the above case. In the process of performing speaker identification with these models, our privacy criteria are:

1. The server should not observe the speech data belonging to the client.
2. The client should not observe the speaker models belonging to the server.

The first criterion is follows directly from the discussion above, the server being able to observe the speech data causes the violation of client privacy. The client can also include in its privacy policy that the user privacy will be protected because if an agency needs to perform surveillance, it will do so using a privacy-preserving speaker identification system.

The second criterion is important because it is possible to identify the speaker by reverse-engineering the speaker models. The client might do this to gain information about the blacklisted individuals the server is performing surveillance on. This would cause problems in the investigation process as the client can convey this information to those individuals.

We consider the adversarial behaviors of the client and the server below. We assume that the client and the server are computationally bounded, so that they cannot directly break encryption and hash functions without the necessary keys. In the speaker identification task, the server tries to gain as much information as possible from the input provided by the client. We assume that the server is legally constrained from using the models for the speakers it has no legal sanction to perform surveillance on. The server cannot do much to disrupt the protocol, the server can use incorrect speaker models, but that would only result in incorrectly identified speaker. As that is not in the interest of the server, we assume that the server is *semi-honest*. Similarly, the client tries to gain information about the server models from the intermediate steps of the protocol. As discussed above, we assume that the client cooperates in the speaker identification task by submitting the correct input, and therefore we also require the client to be *semi-honest*.

7.2.1 Collusion

In speaker identification, there are two types of parties, the client and the server. One client, however, participates in speaker identification tasks with multiple servers and one server participates with multiple clients. There is no motivation for the client and server participating in a speaker identification session to collude as that would only result in the server obtaining the speech input belonging to the client and the client obtaining the speaker models belonging to the server, both direct violations of the privacy criteria discussed above. There is also no motivation for the clients to collude amongst themselves or the servers to collude among themselves, as the clients and the servers only interact with each other individually.

Similar to the case of collusion in privacy-preserving speaker verification (Sect. 4.2.2), the client can collude with one server to gain information about other clients, and the server can collude with one client to gain information about other servers. This, however, is also not possible because the speech input and the speaker models are obfuscated: encrypted or hashed by the client and the server that own these respectively using their keys or salts.

7.2.2 Information Leakage After Multiple Interactions

In a speaker identification, each interaction between the client and the server is fresh. In case of GMM-based protocols, the server or the client can simply use a different encryption key in each execution and in case of LSH-based protocols, the server and the client can use different salts. In this way, either party cannot obtain new information about the other party after multiple executions of the protocol.

Chapter 8
Privacy-Preserving Speaker Identification Using Gaussian Mixture Models

8.1 Introduction

In this chapter we present a framework for privacy-preserving speaker identification using Gaussian mixture models (GMMs). As discussed in the previous chapter, we consider two parties, the client having the test speech sample, and the server having a set of speaker models who is interested in performing the identification. Our privacy constraints are that the server should not be able to observe the speech sample and the client should not be able to observe the speaker models.

As speaker identification can be considered as a generalization of speaker verification to multiple speakers, the protocols for the two problems are similar in principle, as we are evaluating GMMs in both cases. The main difference between the two problem settings is in how the speaker models are stored; in verification, the system stores encrypted speaker models, in identification, the server has plaintext models. This allows us to construct two variants of the speaker identification framework, where the client sends the encrypted speech data to the server or the server sends encrypted models to the client. We discuss the advantages of the two frameworks later in the chapter. Another difference between the protocols for speaker identification and speaker verification is in final maximum computation step. In speaker verification, the user and the system needs to compare only two encrypted numbers which can be done using the millionaire protocol. In speaker identification, the server or the client needs to compare K encrypted numbers, for which we construct the private maximum computation protocol.

The rest of the chapter proceeds as follows: we formally discuss the architecture for the privacy-preserving speaker identification system in Sect. 8.2, and present the identification protocols using homomorphic encryption in Sect. 8.3. We present experiments on a prototype implementation of the protocol in Sect. 8.4.

M. A. Pathak, *Privacy-Preserving Machine Learning for Speech Processing*,
Springer Theses, DOI: 10.1007/978-1-4614-4639-2_8,
© Springer Science+Business Media New York 2013

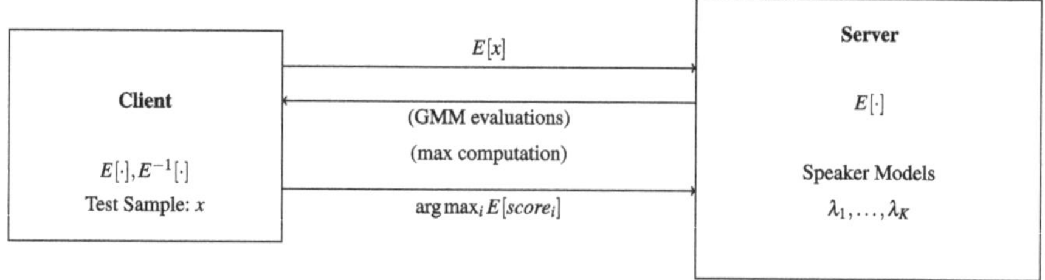

Fig. 8.1 GMM-based speaker identification: client sends encrypted speech sample to the server

8.2 System Architecture

We assume that the server knows the set of speakers $\{S_1, \ldots, S_K\}$ that it is interested in identifying and has data for each speaker. This data could be publicly available or extracted by the server in its previous interactions with the speakers. The server uses a GMM to represent each speaker. The server also trains a universal background model (UBM) λ_U over combined data from a diverse set of speakers. We consider the UBM to represent the none of the above case, where the test speaker is outside the set $\{S_1, \ldots, S_K\}$. The server obtains GMMs for individual speakers $\{\lambda_1, \ldots, \lambda_K\}$ by either training directly over the data for that speaker or by performing MAP adaptation with the UBM.

The client has access to the speech sample, that it represents using MFCC features. To perform identification, the server needs to evaluate the $K + 1$ GMMs $\{\lambda_1, \ldots, \lambda_K, \lambda_U\}$ over the speech sample. The server assigns the speaker to the GMM that has the highest probability score, $\arg\max_i P(x|\lambda_i)$.

In Fig. 8.1 we design the first variant of the speaker identification framework in which the client sends speech samples to the server. Initially, the client generates a public/private key pair $E[\cdot]$, $E^{-1}[\cdot]$ for the homomorphic cryptosystem and sends the public key to the server. The client will then encrypt the speech sample and send it to the server. The server uses the GMM evaluation protocol for each speaker model to obtain the $K + 1$ encrypted probability scores. The server and client then engage in the private maximum computation protocol where only the server knows the model having the highest score at the end.

In the second variant of the speaker identification framework denoted in Fig. 8.2, the server sends models to the client. To do this privately, the server as opposed to the client creates a public/private key pair $E[\cdot]$, $E^{-1}[\cdot]$ for the homomorphic cryptosystem and sends the public key to the client. The server encrypts all GMMs using this key and sends it to the client. The client evaluates all the GMMs over the speech sample it has and obtains $K + 1$ encrypted scores. The client and the server then participate in the private maximum computation protocol where only the server will know the model having maximum score at the end.

We present the cryptographic protocols for the two variants of the privacy-preserving speaker identification framework in the next section.

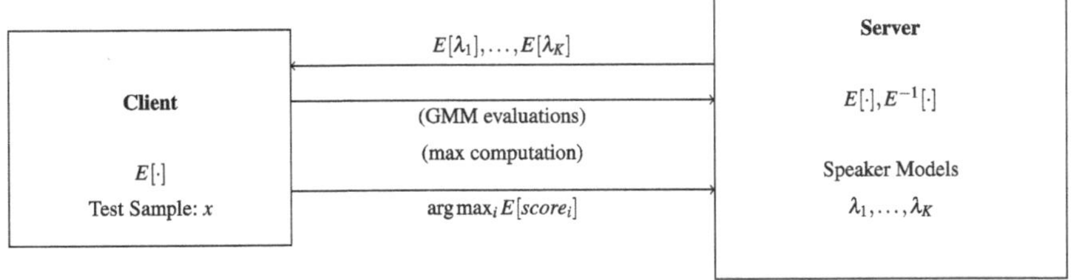

Fig. 8.2 GMM-based speaker identification: server sends encrypted models to the client

8.3 Speaker Identification Protocols

As discussed above, the speaker identification task involves GMM evaluation. The private GMM evaluation protocols here are similar to those used in speaker verification (Sect. 5.2); we modify the protocols to fit the system architecture for speaker identification. We reuse the W construction for representing a Gaussian as given by Eq. 5.2:

$$\log P(x|i) = \log \mathcal{N}(x; \mu, \Sigma) = x^T W.$$

We construct the following GMM evaluation protocols for the two cases: evaluating over private speech data and evaluating over private speaker models and later use the protocols in the speaker identification protocols.

8.3.1 Case 1: Client Sends Encrypted Speech Sample to the Server

GMM Evaluation Protocol with Private Speech Sample.
Inputs:

(a) Client has the test sample x with frame vectors $\{x_1, \ldots, x_T\}$ and both encryption key $E[\cdot]$ and decryption key $E^{-1}[\cdot]$.
(b) Server has the GMM λ with N mixture components $W_1, \ldots, W_N$ and the encryption key $E[\cdot]$.

Output: Server obtains the encrypted score $E[\log P(x|\lambda)]$.

1. The client encrypts the frame vectors $E[x_t]$ and sends it to the server.
2. For each Gaussian matrix W_j and each frame vector $E[x_t] = (E[x_{t1}], \ldots, E[x_{tn}])$, the server computes the inner product $E[x_t^T W_j]$ homomorphically as

$$\prod E[x_{ti}]^{W_{ji}} = E\left[\sum x_{ti} W_{ji}\right] = E[x_t^T W_j].$$

3. The server and the client then participate in the *logsum protocol* (Sect. 5.5) to obtain $E[\log P(x_t|\lambda)] = E[\log \sum_j e^{x_t^T W_j}]$.
4. The server adds the logsums homomorphically to obtain the $E[\log P(x|\lambda)]$.

By using this protocol, the server is able to privately evaluate a GMM it has in plaintext over speech data belonging to the client. As the server does not have the private key, it is not able to observe the encrypted speech sample provided by the client and the final encrypted probability score. The server executes this protocol for all the $K + 1$ GMMs $\{\lambda_1, \ldots, \lambda_K\}$, including the UBM and obtains the encrypted scores $E[\log P(x|\lambda_1)], \ldots, E[\log P(x|\lambda_K)]$ and the server needs to find the GMM having the maximum probability score.

The server and the client participate in the following private maximum computation protocol for this purpose. Our construction is based on the SMAX protocol of Smaragdis and Shashanka (2007) and the blind and permute protocol of Atallah et al. (2003). The basic idea behind blind and permute is that given two number pairs x_1, x_2 and y_1, y_2, comparing the difference $x_1 - x_2 \geq y_2 - y_1$ implies $x_1 + y_1 \geq x_2 + y_2$. The server and the client additively share the encrypted scores and compute the pairwise differences between their respective shares, and then privately compare the entries to find the maximum score. Prior to this, the server permutes the encrypted scores in order to prevent the client from knowing the true index of the maximum score.

Private Maximum Computation Protocol.
Inputs:

(a) Client both encryption key $E[\cdot]$ and decryption key $E^{-1}[\cdot]$.
(b) Server has $K + 1$ encrypted numbers $\{E[a_1], \ldots, E[a_{K+1}]\}$.

Output: Server knows the index of the maximum number, $\arg\max_i a_i$.

1. The server applies a random permutation π to the encrypted numbers to change their sequence into $\{E[a_{i_1}], \ldots, E[a_{i_{K+1}}]\}$.
2. The server generates $K + 1$ random numbers $\{r_1, \ldots, r_{K+1}\}$ and subtracts them homomorphically from the encrypted numbers to get $\{E[a_{i_1} - r_1], \ldots, E[a_{i_{K+1}} - r_{K+1}]\}$. The server sends these numbers to the client.
3. The client decrypts the numbers to obtain $\{a_{i_1} - r_1, \ldots, a_{i_{K+1}} - r_{K+1}\}$ which we denote by the sequence $\{s_1, \ldots, s_{K+1}\}$. The client computes pairwise differences between each of these numbers: $s_1 - s_2, \ldots, s_K - s_{K+1}$.
4. The server similarly computes the pairwise differences $r_1 - r_2, \ldots, r_K - r_{K+1}$. The client and the server execute the secure comparison protocol (e.g., using Millionaire protocol) for each difference entry.
5. At the end of the computation, both the client and the server will know that the entry $r_j + s_j = a_{i_j}$ is the largest. The server will reverse the permutation π to obtain the corresponding index of a_{i_j} in the original sequence.

The above protocol allows the server to privately find the GMM having the maximum score which corresponds to the identified speaker. Both the server and the client do not observe the scores, as that will also reveal additional information about the speech sample or the GMMs. The additive randomization r_i hides the scores and the perturbation π hides the order of the GMMs from the client. The latter is important because it does not reveal which speaker was identified by the server to the client.

8.3.2 Case 2: Server Sends Encrypted Speaker Models to the Client

GMM Evaluation Protocol with Private Speaker Model.
Inputs:

(a) Client has the test sample x with frame vectors $\{x_1, \ldots, x_T\}$ and the encryption key $E[\cdot]$.
(b) Server has the GMM λ with N mixture components $W_1, \ldots, W_N$ and both encryption key $E[\cdot]$ and decryption key $E^{-1}[\cdot]$.

Output: Client obtains the encrypted score $E[\log P(x|\lambda)]$.

1. The server encrypts the Gaussian matrices $\{E[W_1], \ldots, E[W_N]\}$ and sends it to the client.
2. For each frame vector x_t and each encrypted Gaussian matrix, the client computes the inner product $E[x_t^T W_j]$ homomorphically as

$$\prod E[W_{ti}]^{x_{ji}} = E\left[\sum W_{ti} x_{ji}\right] = E[x_t^T W_j].$$

3. The client and the server then participate in the *logsum protocol* (Sect. 5.5) to obtain $E[\log P(x_t|\lambda)] = E[\log \sum_j e^{x_t^T W_j}]$.
4. The client adds the logsums homomorphically to obtain the $E[\log P(x|\lambda)]$.

The client uses the above protocol to evaluate $K+1$ GMMs provided by the server in ciphertext on its speech data. The client obtains probability scores encrypted by the server at the end of the protocol. To perform speaker identification, only the server needs to find the GMM having the maximum score. The client cannot transfer the scores to the server, as that would lead to the loss of privacy of the client speech data. We instead construct the following protocol that uses another set of keys generated by the client and then reuse the private maximum computation protocol we discussed above. We cannot directly use the private maximum computation protocol with the parties reversed, i.e., the client as the server and the server as the client. This is because our privacy criteria require that only the server should know about the identified speaker and for that the server needs to perform the random permutation.

Client-Initiated Private Maximum Computation Protocol.
Inputs:

(a) Client has $K + 1$ encrypted numbers $\{E[a_1], \ldots, E[a_{K+1}]\}$.
(b) Server has both encryption key $E[\cdot]$ and decryption key $E^{-1}[\cdot]$.

Output: Server knows the index of the maximum number, $\arg\max_i a_i$.

1. The client generates a new encryption key $E_1[\cdot]$ and a decryption key $E_1^{-1}[\cdot]$, and sends the encryption key to the server.
2. The client generates $K + 1$ random numbers $\{r_1, \ldots, r_{K+1}\}$ and subtracts them homomorphically from the encrypted numbers to get $\{E[a_1 - r_1], \ldots, E[a_{K+1} - r_{K+1}]\}$. The client sends these ciphertexts to the server.
3. The client encrypts the random numbers $\{r_1, \ldots, r_{K+1}\}$ using $E_1[\cdot]$ and sends the ciphertexts $\{E_1[r_1], \ldots, E_1[r_{K+1}]\}$ to the server.
4. The server decrypts the first set of ciphertexts using $E^{-1}[\cdot]$ to obtain $\{a_1 - r_1, \ldots, a_{K+1} - r_{K+1}\}$. The server re-encrypts this set using $E_1[\cdot]$ to get $\{E_1[a_1 - r_1], \ldots, E_1[a_{K+1} - r_{K+1}]\}$.
5. The server homomorphically adds the two sets of ciphertexts under $E_1[\cdot]$ to get $\{E_1[a_1], \ldots, E_1[a_{K+1}]\}$. The server uses this as input to the original private maximum computation protocol with the client to obtain the index of the maximum number, $\arg\max_i a_i$.

Comparison of the Two System Configurations

In the first variant of the system architecture described in Case 1, the main computation and communication overhead is due to the client encrypting its speech sample. This overhead is directly proportional to the length of the sample T, with 100 frames per second. In the remainder of the GMM evaluation and private maximum computation protocols, the client and server exchange small vectors that are independent of the sample length.

In the second variant described in Case 2, the main overhead is due to the server encrypting its speaker models. As discussed above, we represent the speaker model using N matrices $W_1, \ldots, W_N$ representing a single mixture component. Each W_j matrix is of size $(d + 1) \times (d + 1)$, where d is the dimensionality of the frame vector. In our implementation, we use $d = 39$ with MFCC features with $N = 32$. This size is independent of the sample length. The client evaluates these models on its own unencrypted speech data. Similar to the first variant, the overhead from the remainder of the private computation is relatively small.

In this way, the cost of using the two configurations is dependent on the length of the speech sample. If the length T is typically smaller than the matrix size $N \times (d + 1) \times (d + 1)$, it is advantageous to use the first variant, where the client encrypts the speech sample. If T is larger than the product, it is advantageous to use the second variant. As compared to speaker verification, speaker identification is performed on relatively large amount of speech input, often multiple minutes long in practical scenarios such as surveillance. In these situations, it is significantly more efficient to use the second variant. On the other hand the speech input can be only a few seconds long in problems where speaker identification is used as an initial step for other speech processing tasks. In these situations, it is efficient to use the first variant.

8.4 Experiments

We present the results of experiments on the privacy preserving speaker identification protocols described above. Since our basic adaptation algorithm itself is the same as that in Bimbot et al. (2004) and can be expected to be more or less as accurate as it, the key aspect we need to evaluate is the computational overhead of the proposed protocols. We created a prototype implementation of the verification protocol in C++ and used the variable precision arithmetic libraries provided by OpenSSL (2010) to implement the Paillier cryptosystem. We performed the experiments on a 2 GHz Intel Core 2 Duo machine with 3 GB RAM running 64-bit Ubuntu.

8.4.1 Precision

Both variants of the speaker identification protocol achieved the same final probability scores as the non-private algorithm up to 5 digits of precision.

8.4.2 Accuracy

We used the YOHO dataset (Campbell 1995) to measure the accuracy of the speaker identification task. We used the experimental setup similar to Reynolds and Rose (1995), Michalevsky et al. (2011). We observed the same accuracy for the two variants of the speaker identification protocol as they both result in the same speaker scores.

We trained a UBM on a random subset of the enrollment data and performed MAP adaptation with the enrollment data for K speakers to obtain the speaker models. We evaluate the UBM and the speaker models on the test data for the K speakers, in addition to the speakers outside the set representing the none of the above case. We used identification accuracy, i.e., the fraction of the number of times a test speaker was identified correctly.

$$\text{Accuracy} = \frac{\#(\text{Identified Speaker} = \text{Correct Speaker})}{\#(\text{trials})}.$$

For a 10-speaker classification task, we observed 87.4 % accuracy.

8.4.3 Execution Time

We measured the execution times for the verification protocols using Paillier encryption keys of sizes 256 and 1024-bits. We identify a 1 s speech sample containing 100 frame vectors using $K + 1$ speaker models each containing 32 mixture components using the two variants of the speaker identification protocol. We report the time for

Table 8.1 GMM-based speaker identification: execution time Case 1: client sends encrypted speech sample to the server

Steps	256-bit	1024-bit
Encrypting $x_1, \ldots, x_T$ (s)	136.64	7348.32
Evaluating speaker model (s)	95.24	1189.98
Total for 10 speakers	1089.04 s	19248.12 s
	$\sim$18 min, 9 s	$\sim$5 h, 20 min

Table 8.2 GMM-based speaker identification: execution time Case 2: server sends encrypted speaker models to the client

Steps	256-bit	1024-bit
Encrypting $W_1, \ldots, W_N$ (s)	32.17	1854.07
Evaluating speaker model (s)	460.07	4796.26
Total for 10 speakers	4922.40 s	66503.30 s
	$\sim$1 h, 22 min	$\sim$18 h, 28 min

evaluating 10 speaker models. As the time required for evaluating each model is approximately the same, these numbers can be appropriately scaled to obtain estimated execution time for more speakers.

We summarize the results in Table 8.1 and 8.2.

8.5 Conclusion

In this chapter, we developed a framework for privacy-preserving speaker identification using GMMs and the Paillier cryptosystem. In this model, the server is able to identify which of the K speakers best corresponding to the speech input provided by the client without being able to observe the input. We present two variants of the framework, where either the client submits encrypted speech to the server or the server submits encrypted speaker models to the client.

We perform accuracy and timing experiments on a prototype implementation of the protocol. On a 10-speaker identification task, we observe 87.4 % accuracy. For identifying a 1 s speech input on the same task, the first variant of the protocol required 18 min and the second variant of the protocol required 1 h, 22 min.

References

Atallah MJ, Kerschbaum F, Du W (2003) Secure and private sequence comparisons. In: Workshop on privacy in the electronic society, pp 39–44

Bimbot F, Bonastre J-F, Fredoille C, Gravier G, Ivan M-C, Sylvain M, Teva M, Javier O-G, Dijana P-D, Douglas R (2004) A tutorial on text-independent speaker verification. EURASIP J Appl Signal Process 4:430–451

Campbell JP (1995) Testing with the YOHO CD-ROM voice verification corpus. In: IEEE international conference on acoustics, speech and signal processing, pp 341–344
Michalevsky Y, Talmon R, Cohen I (2011) Speaker identification using diffusion maps. In: European signal processing conference
OpenSSL. http://www.openssl.org/docs/crypto/bn.html
Reynolds DA, Rose RC (1995) Robust text-independent speaker identification using gaussian mixture speaker models. IEEE Trans Speech Audio Process 3(1):72–83
Paris S, Madhusudana S (2007) A framework for secure speech recognition. IEEE Trans Audio Speech Lang Process 15(4):1404–1413

Chapter 9
Privacy-Preserving Speaker Identification as String Comparison

9.1 Introduction

In this chapter, we present a framework for speaker identification using locality sensitive hashing (LSH) and supervectors. Instead of GMM evaluation that we considered in the previous chapter, we transform the problem of speaker identification into string comparison. The motivation behind doing so is to perform speaker identification with privacy while having minimal computational overhead. Similar to the speaker verification framework of Chap. 6, we convert the utterances into supervector features (Campbell et al. 2006) that are invariant with the length of the utterance. By applying the LSH transformation to the supervectors, we reduce the problem of nearest-neighbor classification into string comparison. As LSH is not privacy-preserving, we apply the cryptographic hash function to the LSH keys. This allows us to check if two LSH keys match without knowing their contents.

The main difference from this speaker identification framework and the framework for speaker verification of Chap. 6 is in its application setting. Here, we assume that the client has the test speech sample and the server already has speech data from multiple speakers. The client and server need to compute the supervector over their speech data, apply LSH functions to get the LSH keys and finally the cryptographic hash function. The client transfers the cryptographic hashes to the server who then matches them to the hashes derived from the speech data. One issue with this approach is that the hashes need to be salted to prevent the server from performing a brute-force attack to recover the original LSH keys. It is also not possible to keep the salt private with the client, as we did in the speaker verification framework, as the server also needs to compute salted hash over its own data. Our privacy constraints also require that the client should not observe the speech data belonging to the server, so the server cannot simply transfer its supervectors to the client for hashing. To solve this problem, we construct a oblivious salting protocol, where the client and the server are able to compute salted hashes without either party knowing the contents of the salt. We present the detailed protocol construction in Sect. 9.3.

Similar to speaker verification with LSH, our main goal is not to develop speaker identification algorithms that achieve higher accuracy as compared to Gaussian

M. A. Pathak, *Privacy-Preserving Machine Learning for Speech Processing*, Springer Theses, DOI: 10.1007/978-1-4614-4639-2_9,

mixture models (GMMs) based algorithms. We are principally interested performing speaker identification while satisfying the same privacy constraints but with a minimal computational overhead, while achieving feasible accuracy. We present the results of experiments with the LSH based speaker identification framework using the LSH functions defined over Euclidean and cosine distances in Sect. 9.4.

9.2 System Architecture

We assume that the server knows the set of speakers $\{S_1, \ldots, S_K\}$ that it is interested in identifying and has data in the form of n speech samples for each speaker. This data could be publicly available or extracted by the server in its previous interactions with the speakers. The server trains a universal background model (UBM) λ_U over combined data from a diverse set of speakers. We consider the UBM to represent the none of the above case, where the test speaker is outside the set $\{S_1, \ldots, S_K\}$. The server performs MAP adaptation of λ_U with each speech sample belonging to each speaker to obtain the corresponding supervectors as described in Sect. 2.2.3. The server therefore has n supervectors for each of the K speakers. The server also obtains n speech samples from a set of imposter speakers to model the none of the above case, where the test sample does not match any of the K speakers; this data could be from the training data of the UBM. The server also performs MAP adaptation for this data to obtain the corresponding supervectors, considering it as the $(K + 1)$th speaker.

The server generates l LSH functions $L_1(\cdot), \ldots, L_l(\cdot)$, each of length k-bits and applies them to each supervector to obtain l keys for each. The $l \times n$ matrix of keys represents the server data for one speaker. The server also makes the UBM λ_U and the LSH function public to all clients. We assume a client to have a test speech sample. The client obtains λ_U from the server and performs MAP adaptation to obtain the corresponding supervector. The client also applies the LSH functions to obtain l keys. For each of the K speakers, the server needs to count the number of training supervectors are matched by the test supervector, where we consider two supervectors as matched if at least one LSH key is common between them. The server identifies the speaker by the one having the most number of matches on the training supervectors.

As the LSH keys can be utilized to obtain private information about the speaker, the client and the server need to apply a salted cryptographic hash function $H[\cdot]$ to the LSH keys computed on the test supervector. As mentioned above, we construct an oblivious salting protocol for the client and the server to compute a salted cryptographic hash together without any party knowing the salt. Using this protocol, the client and the server compute salted hashes over their LSH keys and the server performs the matching to identify the speaker. We present an overview of the supervector-based protocol in Fig. 9.1.

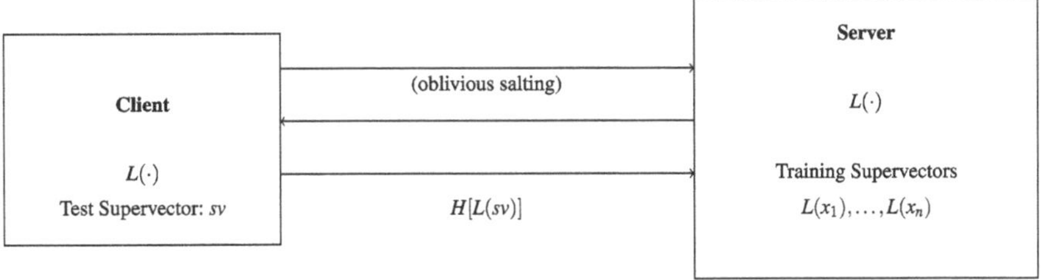

Fig. 9.1 Supervector-based speaker identification protocol

9.3 Protocols

9.3.1 Oblivious Salting

We first present the oblivious salting protocol that enables the client and the server to compute the salted hashes without knowing the salt; the construction is similar to (Cristofaro and Tsudik 2010). The key idea behind this protocol is the use of two random oracles $H : \{0, 1\}^* \mapsto: \{0, 1\}^\kappa$ for a security parameter κ and $H' : \{0, 1\}^* \mapsto \mathbb{Z}_q^*$. We can model H using any general hash function, such as SHA-2, but H' needs special consideration due to its range. One method to model H' is as the hash in the group, $H'[x] = x^{(p-1)/q} \mod q$ where $q \mid p - 1$ and p is a large prime number. One advantage of H' is that it allows salting by modular exponentiation: it is infeasible to recover $H'[x]$ from $H'[x]^t \mod q$ if t is unknown. The client and server respectively generate random numbers r and s for this purpose.

The client and the server also generate random numbers r and s in the finite field $\mathbb{Z}_q$ and exchange g^r and g^s with each other, similar to Diffie-Hellman key exchange (Diffie and Hellman 1976). Due to the discrete log problem, the client is unable to obtain s from g^s that it obtains from the server, and similarly the server is unable to obtain r from g^r, thereby keeping r and s secret from each other. The client and the server individually use their secret numbers to compute the shared key g^{rs}: the client computes $(g^s)^r = g^{rs}$ and the server computes $(g^r)^s = g^{rs}$. Both the client and the server use this as a shared salt in the hash function H. As we discuss later, the parties use this protocol with their LSH keys as input.

Protocol for hashing with oblivious salt.
 Inputs:

(a) Client has a.
(b) Server has b.
(c) Common Input: Prime numbers p, q such that $q \mid p-1$, group $\mathbb{Z}_q$ with generator g, and two hash functions H and H'.

Output: Client has $\{H[a \parallel z]$ and server has $\{H[b \parallel z]$ for an oblivious salt z.

1. The server generates a random number $s \in \mathbb{Z}_q$ and sends g^s to the client.

2. The client generates a pair of random numbers $r, r' \in \mathbb{Z}_q$. The client uses the server's input to compute g^{rs} and $g^{r's}$.
3. The client applies the hash function H' over its input to obtain $H'[a]$, and uses r' to compute $H'[a]g^{r'}$. The client sends this to the server.
4. The server exponentiates this by s to obtain $H'[a]^s g^{r's}$ and sends it back to the client.
5. The client multiplies this term by $g^{-r's}g^{rs}$ to obtain $H'[a]^s g^{rs}$.
6. The client applies the hash function H to obtain $H[H'[a]^s g^{rs}]$. We consider the terms g^{rs} and s together as the salt z, and denote the hashed term by $H[a \parallel z]$.
7. The server applies the hash function H' over its input to obtain $H'[b]$, exponentiates it by s, and uses s' to compute $H'[b]^s g^{s'}$. The server sends this to the client.
8. The client multiplies this with g^{rs} to obtain $H'[b]^s g^{s'} g^{rs}$, which the client sends back to the server.
9. The server multiplies this with $g^{-s'}$ to obtain $H'[b]^s g^{rs}$. The server then applies the hash function H to obtain $H[H'[b]^s g^{rs}]$, which is equal to $H[b \parallel z]$ for the same salt z.

In the intermediate steps of the protocol, the client and the server multiply their hashes $H'[a]$ and $H'[b]$ with $g^{r'}$ and $g^{s'}$ respectively. This operation protects the hash from the other party by effectively salting it with r' and s'; the server does not know r' and the client does not know s'. The client and server, therefore, do not observe each others inputs. The client and the server are able to reverse their respective salts by multiplying the output by $g^{-r'}$ and $g^{-s'}$.

The final salted hash that we obtain at the end of the protocol has the form $H[x \parallel z] = H[H'[x]^s g^{rs}]$ for the input element x. The salt is a function of s and r, which are respectively known to the client and the server. The client or the server, therefore, cannot compute the hash on their own as the client does not know s and the server does not know g^{rs}. If the server intends to use a brute-force strategy to break this hash, i.e., identify a given $H[H'[a]^s g^{rs}]$, it would need to compare it with the hashes computed over all possible values of r, which would be infeasible due to the enormously large number of possible values that $r \in \mathbb{Z}_q$ can take.

This protocol requires transferring 4 data transfers beyond the initial transfer of g^s. We can easily generalize the oblivious salting protocol to case where the client and the server have a set of inputs instead of a single element. The client can apply H' to the entire input set and transfer $\{H'[a_1]g^{r'}, \ldots, H'[a_m]g^{r'}\}$ to the server, and similarly obtain $\{H'[a_1]^s g^{r's}, \ldots, H'[a_m]^s g^{r's}\}$. In this way, the protocol will require $2m + 2n$ data transfers.

9.3.2 Speaker Identification

We present the speaker identification protocol using LSH keys hashed with the oblivious salting protocol. We show only one LSH function for simplicity; in practice we use l LSH functions.

Speaker Identification Protocol
Inputs:

(a) Client has the test speech sample x.
(b) Server has n speech samples for $K + 1$ speakers and the UBM λ_U.

Output: Server knows the identified speaker.

1. The server generates the LSH function $L(\cdot)$ and sends it to the client.
2. The client and the server perform MAP adaptation of each of their speech samples with λ_U to obtain the compute respective supervectors.
3. The client and the server then apply LSH function to obtain the keys: $L(sv)$ for the client and $\{L(sv_{11}), \ldots, L(sv_{n(K+1)})\}$ for the server. The client and the server apply $L(\cdot)$ to their supervectors to obtain the LSH keys.
4. The client and the server execute the oblivious salting protocol to compute salted hashes over all of their LSH keys: $H[L(sv)]$ and $\{H[L(sv_{11})], \ldots, H[L(sv_{n(K+1)})]\}$.
5. The client sends its hashed LSH key $H[L(sv)]$ to the server.
6. The server matches the hash obtained from the client to the hashes for each of the $K + 1$ speakers. The server identifies the speaker by choosing the one with most matches.

In this protocol, the server only observes the hash that is submitted by the client. As the hash is salted using the oblivious salting protocol, the server is not able to use a brute-force strategy to feasibly identify the underlying LSH key. The client does not observe anything about the server's data and the output, as the server performs the matching locally. For efficiency, the server can pre-compute the supervectors and the LSH keys in multiple executions of the protocol. In case, the server is executing the protocol with the same client using a different test sample, the hashes also do not need to be recomputed.

9.4 Experiments

We experimentally evaluate the privacy preserving speaker identification protocols described above for accuracy and execution time. We perform the experiments on the YOHO dataset (Campbell 1995). We perform all the execution time experiments on a laptop running 64-bit Ubuntu 11.04 with 2 GHz Intel Core 2 Duo processor and 3 GB RAM.

9.4.1 Accuracy

We used Mel-frequency cepstral coefficient (MFCC) features augmented by differences and double differences, i.e., a recording x consists of a sequence of

Table 9.1 Average accuracy for the three LSH strategies: Euclidean, cosine, and combined (Euclidean & cosine)

Euclidean	Cosine	Combined
79.5 %	78 %	81.3 %

39-dimensional feature vectors $x_1, \ldots, x_T$. We trained a UBM with 64 Gaussian mixture components on a random subset of the training data belonging to all users. We obtained the supervectors by individually adapting all enrollment and verification utterances to the UBM.

Similar to Chap. 8, we use identification accuracy as the evaluation metric. We observed that the highest accuracy was achieved by using $l = 200$ instances of LSH functions each of length $k = 20$ for both Euclidean and cosine distances. A test sample considered to match a training sample if at least one of their LSH keys matches. The score for a test sample with respect to a speaker is given by the number of training samples for that speaker it matched. We also consider supervectors derived from randomly chosen imposter samples to represent the none of the above case.

We report the average EER for different speakers in Table 9.1. Similar to the speaker verification experiments in Chap. 6, we observe that LSH for Euclidean distance performs better than LSH for cosine distance. We also used the combined LSH keys for Euclidean and cosine distances and found that this strategy performed the best.

9.4.2 Execution Time

As compared to a non-private variant of a speaker identification system based on supervectors, the only computational overhead is in applying the LSH and the hash function. For a $64 \times 39 = 2496$-dimensional supervector representing a single utterance, the computation for both Euclidean and cosine LSH involves a multiplication with a random matrix of size 20×2496 which requires a fraction of a millisecond. Performing this operation 200 times required 15.8 ms on average.

With respect to privacy, LSH-based speaker verification has minimal overhead as compared to secure multiparty computation (SMC) protocols using homomorphic encryption. This is primarily due to the fact that cryptographic hash functions such as SHA-256 are orders of magnitude faster than homomorphic encryption. In LSH-based speaker identification, however, we cannot directly use SHA-256 due to the need for oblivious salting. As we have seen above, the protocol for hashing with oblivious salt requires modular exponentiation and is therefore comparatively less efficient as cryptographic hash functions.

We measure the execution time required for hashing a pair of numbers. We created a prototype implementation of the oblivious salting protocol in C++ and used the variable precision arithmetic libraries provided by OpenSSL (2010) to implement the modular arithmetic operations. We used the prime number p of size 1024-bits as that is considered to provide state of the art security.

For hashing one key represented as an integer, we observed that on average the client requires 14.71 ms and the server requires 8.48 ms. The discrepancy is due to the fact that we need to perform more operation when hashing client's input. We represent each speech sample by its supervector and then 200 LSH keys. Hashing 200 elements would require 2.942 s for the client and 1.696 s for the server. Beyond this, the identification protocol only consists of exactly matching the 1024-bit long hashed keys derived from the test sample to those obtained from the training data.

Although this execution time is larger than applying cryptographic hash functions on LSH keys as we do speaker verification, it is significantly shorter than encrypting the speech sample itself using Paillier encryption. In Chap. 8, we saw that encrypting a 1 s speech sample requires 136.64 s.

9.5 Conclusion

In this chapter, we developed an algorithm for speaker identification using supervectors with locality sensitive hashing (LSH). Here, we developed an cryptographic hashing algorithm in which two parties can perform oblivious salting to protect their data from each other. As was the case with speaker verification, the LSH-based protocols have smaller computational overhead as compared to the GMM-based protocols. We observed that the protocol requires 2.9 s for each input supervector. In terms of the accuracy, we observe that the algorithm has 81.3 % accuracy, which is comparatively lower than the GMM-based algorithm. Again, these two algorithms represent a trade-off between execution time and accuracy.

References

Campbell JP (1995) Testing with the YOHO CD-ROM voice verification corpus. In: IEEE international conference on acoustics, speech and signal processing, pp 341–344

Campbell WM, Sturim DE, Reynolds DA, Solomonoff A (2006) SVM based speaker verification using a GMM supervector kernel and NAP variability compensation. In: IEEE international conference on acoustics, speech and signal processing

De Cristofaro E, Tsudik G (2010) Practical private set intersection protocols with linear complexity. In: Financial cryptography

Diffie W, Hellman ME (1976) New directions in cryptography. IEEE Trans Inform Theor IT-22(6):644–654

OpenSSL (2010) http://www.openssl.org/docs/crypto/bn.html

Part IV

Privacy-Preserving Speech Recognition

Chapter 10
Overview of Speech Recognition with Privacy

10.1 Introduction

Speech recognition is the task of identifying the spoken words from a given speech sample. The speech recognition system consists of an acoustic model that attempts to match the speech patterns and a language model that attempts to match the underlying linguistic structure of the speech. The acoustic model is typically represented using a probabilistic model such as a hidden Markov model (HMM). In most speech recognition systems, we assume that the entire set of words is typically available beforehand in the form of a dictionary. The language model is represented using a word-based n-gram model.

The simplest form of speech recognition is that over isolated words. We assume that the speech is tokenized into words and the system is given speech samples consisting of individual words. Here, we represent each word using a single HMM. The most general form of speech recognition is transcribing continuous speech, i.e., the speech as it is naturally produced by a speaker with intermittent silences and filler sound. This is a significantly more complex task than isolated word recognition and we typically achieve much lower accuracies. Here, the acoustic model consists of a concatenation of multiple HMMs.

As we discussed in Chaps. 5 and 8, there is a significant computational overhead in the homomorphic encryption based protocols for GMM evaluation. This overhead is also applicable in developing similar protocols for HMM inference. This would be greatly amplified in the case of privacy-preserving continuous speech recognition, making the computation infeasible for practical purposes. We, therefore, restrict ourselves to creating an algorithm for privacy-preserving isolated-word recognition.

10.2 Client-Server Model for Speech Recognition

Speech recognition finds widespread applications such as telephony, voice-based control interfaces, spoken dialog systems, speech-to-text input. In many cases, the speech input is procured from a thin-client device such as a smartphone. As the

M. A. Pathak, *Privacy-Preserving Machine Learning for Speech Processing*,
Springer Theses, DOI: 10.1007/978-1-4614-4639-2_10,
© Springer Science+Business Media New York 2013

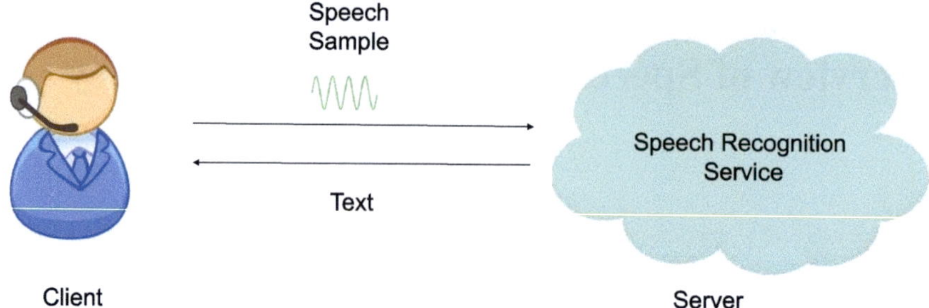

Fig. 10.1 Client-server model for speech recognition

underlying computation involved is fairly intensive and cannot be performed on the client device in real-time, the speech input is transferred to an server that performs the necessary processing and returns the transcribed text as output. As the client may not have the necessary technology in the form of speech recognition algorithms and high quality training data, or the computing infrastructure, this service is hosted on an external platform.

In many cases, such a speech recognition service is hosted on a public cloud computing platform, as this provides multiple benefits including cost reduction and high scalability with respect to demand. A prominent commercial example of such a service is the Google Speech Recognition API. The schematic architecture of a cloud-based speech recognition service is shown in Fig. 10.1.

10.3 Privacy Issues

The client-server model described above has privacy issues dealing with the nature of the user input. The client may not be comfortable sharing the speech sample with an external service if it is a recording of confidential nature, such as financial, legal aspects. The privacy criteria are even more stringent in medical settings due to privacy laws such as HIPAA, that prohibit hospitals from storing patient records in external media.

These privacy issues are usually addressed in the privacy policy of the service provider. The system with the most privacy-sensitive policy, however, requires access to the user input in clear and almost always needs to store it for future analysis and improving its performance. For the scenarios described above, these precautions only address the privacy problems at a superficial level. Even if the client data is stored securely, there are still possible threats to the server storage being breached by an adversary leading to the public disclosure of the client speech data. This is further amplified when the service is hosted on a virtualized storage platform, as there are many potential security flaws that can lead to information being revealed to another party using the same platform (Garfinkel and Rosenblum 2005).

In addition to the server storage being compromised, another source of undesirable publication of client data is from the government regulation about the use of service. The service provider may be required to release the client data due to legal interventions. The terms of the privacy policy usually protect the data till the organization operating the service is functioning. The client data can even be published or sold if the original organization becomes bankrupt.

On the other hand, the server may need to protect its models from the users. This is usually a hard constraint if the server is interested in operating a pay per use service. Additionally, the speech recognition models also form the intellectual property of the server as accurate models are usually obtained from training over high quality speech data which expensive.

In this way, the privacy constraints require that the speech recognition needs to be performed without the server observing the speech input provided by the user, and the client observing the models belonging to the server. In the next section, we look at possible methods by which this could be possible.

10.4 System Architecture

To satisfy the privacy constraints described above, we require the server to operate entirely on obfuscated data submitted by the user. This is an example of secure multiparty computation (SMC), and one commonly used form of obfuscation is encryption. We envision this architecture in Fig. 10.2.

In the next chapter, we develop a protocol that enables the client and the server to collaboratively evaluate the models belonging to the server using clients data while satisfying their privacy constraints. This is achieved using a public-key additively homomorphic cryptosystem. Our framework is computationally asymmetric, where a most of the computation is performed by the server who has larger computational resources at its disposal as compared to the user. which may be running the application on a thin-client device such as a smartphone.

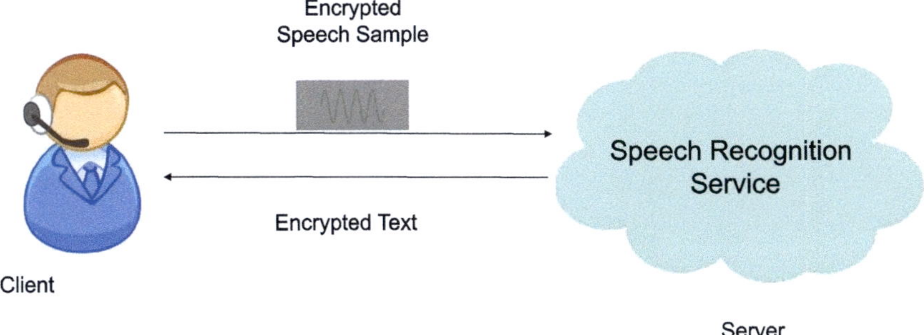

Fig. 10.2 Privacy-preserving client-server model for speech recognition

Reference

Garfinkel T, Rosenblum M (2005) When virtual is harder than real: security challenges in virtual machine based computing environments. In: HotOS, USENIX Association

Chapter 11
Privacy-Preserving Isolated-Word Recognition

11.1 Introduction

In this chapter we present a privacy-preserving framework for isolated word recognition based on Hidden Markov Models (HMMs). Classification based on HMMs is common in machine learning and is nearly ubiquitous in applications related to speech processing. We consider a multi-party scenario in which the data and the HMMs belong to different individuals and cannot be shared. For example, Alice wants to analyze speech data from telephone calls. She outsources the speech recognition task to Bob, who possesses accurate HMMs obtained via extensive training. Alice cannot share the speech data with Bob owing to privacy concerns while Bob cannot disclose the HMM parameters because this might leak valuable information about his own training database.

Probabilistic inference with privacy constraints is, in general, a relatively unexplored area of research. The only detailed treatment of privacy-preserving probabilistic classification appears in Smaragdis and Shashanka (2007). In that work, inference via HMMs is performed on speech signals using existing cryptographic primitives. Specifically, the protocols are based on repeated invocations of privacy-preserving two-party maximization algorithms, in which both parties incur exactly the same protocol overhead. In contrast, we present an asymmetric protocol that is more suitable for client-server interaction; the thin client encrypts the data and provides it to a server which performs most of the computationally intensive tasks. Further, HMM-based probabilistic inference involves probabilities and real numbers which must be approximated for encrypted-domain processing. Accordingly, we consider the effect of finite precision, underflow and exponentiation in the ciphertext domain.

M. A. Pathak, *Privacy-Preserving Machine Learning for Speech Processing,*
Springer Theses, DOI: 10.1007/978-1-4614-4639-2_11,
© Springer Science+Business Media New York 2013

11.2 Protocol for Secure Forward Algorithm

We assume that Alice, a client, sets up a private and public key pair for the Paillier encryption function $E(\cdot)$, and sends only the public key to the server Bob. We assume that, in all calculations involving logarithms, the base of the logarithm is $g \in \mathbb{Z}^*_{N^2}$, which is the parameter used for Paillier encryption.

11.2.1 Secure Logarithm Protocol

Input: Bob has $E(\theta)$
Output: Bob obtains the encryption $E(\log \theta)$ of $\log \theta$, and Alice obtains no information about θ.

1. Bob randomly chooses an integer β, and sends $E(\theta)^\beta = E(\beta\theta)$ to Alice
2. Alice decrypts $\beta\theta$, and then sends $E(\log \beta\theta)$ to Bob.
3. Bob computes $E(\log \beta\theta) \cdot E(-\log \beta) = E(\log \theta + \log \beta) \cdot E(-\log \beta) = E(\log \theta)$

In general $\log \beta$ and $\log \beta\theta$ are not integers, and we resort to integer approximations when we implement the protocol. Specifically, in Step 2, Alice actually sends $E(\lfloor L \log \beta\theta \rfloor)$ to Bob, where L is a large number. For example, with $L = 10^6$, our logarithms are accurate to 6 decimal places. Similarly, in Step 3, Bob actually computes $E(\lfloor L \log \beta\theta \rfloor)E(-\lfloor L \log \beta \rfloor) = E(\lfloor L \log \beta + L \log \theta \rfloor - \lfloor L \log \beta \rfloor)$. Every multiplication by L is compensated by a corresponding division at every decryption step that Alice performs. The effect of these finite precision approximations is further discussed in Sect. 11.3.

11.2.2 Secure Exponent Protocol

Input: Bob has $E(\log \theta)$
Output: Bob obtains the encryption $E(\theta)$, and Alice obtains no information about θ.

1. Bob randomly chooses $\beta \in \mathbb{Z}^*_{N^2}$ and sends $E(\log \theta)E(\log \beta) = E(\log \theta + \log \beta) = E(\log \beta\theta)$ to Alice.
2. Alice decrypts $\log \beta\theta$, and then sends $E(\beta\theta)$ to Bob
3. Bob computes $E(\beta\theta)^{\frac{1}{\beta}} = E(\theta)$ where $\frac{1}{\beta}$ is the multiplicative inverse of β in $\mathbb{Z}^*_{N^2}$.

As before, multiplication by a large number L followed by truncation is used to generate an approximation wherever the logarithm is involved. Thus, in Step 1 of our implementation of this protocol, Bob actually sends $E(\lfloor L \log \theta \rfloor)E(\lfloor L \log \beta \rfloor)$ to Alice. To compensate for the multiplication by L, Alice decrypts the transmission from Bob and divides by L in Step 2.

11.2.3 Secure Logsum Protocol

Input: Bob has $(E(\log\theta_1), E(\log\theta_2), \ldots, E(\log\theta_n))$ and the constant vector $(a_1, a_2, \ldots, a_n)$;
Output: Bob obtains $E(\log\sum_{i=1}^{n} a_i\theta_i)$, and Alice discovers nothing about the θ_i and a_i.

1. With Bob's input $E(\log\theta_1), E(\log\theta_2), \ldots, E(\log\theta_n)$, Alice and Bob execute the Secure Exponent protocol repeatedly so that Bob obtains $E(\theta_1), E(\theta_2), \ldots, E(\theta_n)$.
2. Bob exploits the additive homomorphic property of Paillier encryption to obtain

$$
E\left(\sum_{i=1}^{n} a_i\theta_i\right) = \prod_{i=1}^{n} E(a_i\theta_i) = \prod_{i=1}^{n} E(\theta_i)^{a_i}
$$

3. Bob and Alice execute the Secure Logarithm protocol, at the end of which Bob obtains the encryption $E(\log\sum_{i=1}^{n} a_i\theta_i)$.

11.2.4 Secure Forward Algorithm Protocol

Input: Alice has an observation sequence $x_1, x_2, \ldots, x_T$. Bob has the HMM $\lambda = (A, B, \Pi)$.
Output: Bob obtains $E(\log Pr\{x_1, x_2, \ldots, x_T | \lambda\})$.

We write the matrix B as $[b_1, b_2, \ldots, b_N]$, where for each $j = 1, 2, \ldots, N, b_j$ is a column vector with component $b_j(v_k), k = 1, 2, \ldots, M$. Now, a privacy-preserving version of the forward algorithm for HMMs proceeds as follows:

1. For each $t = 1, 2, \ldots, T$ and $j = 1, 2, \ldots, N$, Bob randomly chooses γ_{tj} and generates a column vector $\log b_j + \gamma_{tj}$.
2. Based on the input x_t, Alice uses 1-of-M OT to obtain $\log b_j(x_t) + \gamma_{tj}$.
3. Alice sends $E(\log b_j(x_t) + \gamma_{tj})$ to Bob
4. Using γ_{tj} and the homomorphic property, Bob computes $E(\log b_j(x_t) + \gamma_{tj}) \cdot E(-\gamma_{tj}) = E(\log b_j(x_t))$ for $j = 1, 2, \ldots, N, t = 1, 2, \ldots, T$.
5. Bob computes $E(\log\alpha_1(j)) = E(\log\pi_j) \cdot E(\log b_j(x_1))$ for $j = 1, 2, \ldots, N$
6. Induction Step: For $j = 1, 2, \ldots, N$, with Bob's input $E(\log\alpha_t(j))$, $j = 1, 2, \ldots, N$ and the transition matrix $A = (a_{ij})$, Alice and Bob run the secure logsum protocol, at the end of which Bob obtains $E(\log\sum_{l=1}^{N} \alpha_t(l)a_{lj})$.
7. For all $1 \leq t \leq T - 1$, Bob computes

$$
E(\log\alpha_{t+1}(j)) = E(\log\sum_{l=1}^{N} \alpha_t(l)a_{lj}) \cdot E(\log b_j(x_{t+1}))
$$

8. Alice and Bob again run a secure LOGSUM protocol, so Bob obtains $E(\log \sum_{j=1}^{N} \alpha_T(j)) = E(\log P(x_1, x_2, \ldots, x_T | \lambda))$

11.2.5 Security Analysis

The secure logarithm, secure exponent and secure logsum protocols rely on multiplicative masking to prevent Alice from discovering θ. Bob cannot discover θ because he does not possess the decryption key for the Paillier cryptosystem. The security of the Forward Algorithm protocol derives from the security of the previous three primitive protocols and the security of 1-of-M oblivious transfer (OT).

11.3 Privacy-Preserving Isolated-Word Recognition

Consider the following scenario for privacy-preserving keyword recognition: Alice has a sampled speech signal, which she converts into T frames, where each frame is represented by a d-dimensional vector of mel-frequency cepstral coefficients (MFCC). Thus Alice possesses x_i, $i = \{1, 2, \ldots, T\}$ where each $x_i \in \mathbb{R}^d$. Typically, $d = 39$ is used in practice. Bob possesses Δ different HMMs, each trained for a single keyword. Alice and Bob will execute a secure protocol, at the end of which, Alice will discover the keyword that is most likely to be contained in her speech sample.

Let a d-dimensional vector y of MFCCs have a multivariate Gaussian distribution, i.e., $b_j(y) = \mathcal{N}(\mu_j, C_j)$, where $j = 1, 2, \ldots, N$ indexes the state of a HMM λ. Let $z = [y^T, 1]^T$. Then, $\log b_j(y) = z^T W_j z$, where

$$W_j = \left[\begin{array}{c|c} -\frac{1}{2}C_j^{-1} & C_j^{-1}\mu_j \\ \hline 0 & w_j \end{array} \right] \in \mathbb{R}^{(d+1)\times(d+1)}$$

and $w_j = \frac{1}{2}\mu_j^T C_j^{-1}\mu_j - \frac{1}{2}\log|C_j^{-1}| - \frac{d}{2}\log 2\pi$. The above treatment considers y to be a single multivariate Gaussian random variable, though an extension to mixture of multivariate Gaussians is also possible. Note that the matrix W_j is available to Bob. Further, note that $\log b_j(y)$ is a linear function of products $z_i z_j$ where $i, j \in \{1, 2, \ldots, d+1\}$. This allows us to simplify the Secure Forward Algorithm Protocol of Sect. 11.2.4 as follows:

11.3.1 Simplified Secure Forward Algorithm

1. For each $t = 1, 2, \ldots, T$, Alice sets $z = [x_t^T, 1]^T$. Alice sends to Bob the encryptions of all $z_i z_j$ with $i, j \in \{1, 2, \ldots, d+1\}$.

2. For each HMM state $j = 1, 2, \ldots, N$, Bob obtains the encryption $E(\log b_j(x_t))$ $= E(z^T W_j z)$ using the additive homomorphic property.
3. Bob executes steps 5–8 from Sect. 11.2.4.

Now, the protocol for secure keyword recognition is as follows.

11.3.2 Protocol for Privacy-Preserving Isolated-Word Recognition

Input: Alice has the MFCC feature vectors $x_1, x_2, \ldots, x_T$ corresponding to her privately owned speech sample. Bob has the database of HMMs $\{\lambda_1, \lambda_2, \ldots, \lambda_\Delta\}$ where each HMM λ_δ corresponds to a single keyword, which is denoted by τ_δ;
Output: Alice obtains $\delta^* = \arg\max_\delta \Pr\{x_1, x_2, \ldots, x_T | \lambda_\delta\}$.

1. Alice and Bob execute the protocol of Sect. 11.3.1 for each HMM λ_δ, at the end of which Bob obtains $E(p_\delta) = E(\log \Pr\{x_1, x_2, \ldots, x_T | \lambda_\delta\})$, $\delta = 1, 2, \ldots, \Delta$.
2. Bob chooses an order-preserving matrix[1] $R = (r_{ij})_{\Delta \times \Delta}$ with random coefficients. Using the additively homomorphic property of Paillier encryption, he computes the element-wise encryption given by $(E(p'_1), \ldots, E(p'_\Delta)) = (E(p_1), \ldots, E(p_\Delta)) \cdot R$. Bob sends the result to Alice.
3. Alice decrypts and obtains $\delta^* = \max_\delta p'_\delta = \max_\delta p_\delta$. This is true because R is order-preserving mapping.
4. Alice and Bob perform a 1-of-Δ OT, and Alice obtains the word τ_{δ^*}.

11.3.3 Computational Complexity

Let us consider the computational overhead of the protocol in terms of the number of keywords (Δ), the number of HMM states (N), the time duration of the observation (T) and the size of each observation (d). Alice has an overhead of $O(d^2 T)$ encryptions at the beginning of the protocol. This initial computational overhead is independent of the number of keywords and the size of the HMM. Bob has an overhead of $O(d^2 N \Delta T)$ ciphertext multiplications and exponentiations in order to determine the $E(\log b_j(x_t))$ terms. Finally, to determine the $E(\alpha_T(j))$ terms, Alice and Bob both have an overhead of $O(N \Delta T)$. Thus Alice's total protocol overhead is $O((d^2 + N\Delta)T)$ while Bob's total overhead is $O(d^2 N \Delta T)$, which means that Bob's overhead grows faster than Alice's.

---

[1] See (Rane and Sun 2010) for more details.

11.3.4 Practical Issues

The Paillier cryptosystem is defined over an integer ring but the speech data consists of real numbers. In our implementation, we used numbers accurate to 6 decimal places as explained in Sect. 11.2. Furthermore, whenever there was a floating point number in the ciphertext exponent—of the form $E(a)^b$—then, the fractional part was truncated, and only the integer portion of b was used in the calculation. These approximations introduce errors in the protocol which propagate during the iterations of the forward algorithm.

Another implementation issue is that the value of $\alpha_t(j)$ gets progressively smaller after each iteration, causing underflow after only a few iterations. This problem would not be encountered if the probabilities were expressed in the form $\log \alpha_t(j)$. However, it is necessary to consider the actual value of $\alpha_t(j)$ as seen in Sect. 11.2.4. The underflow problem is resolved by normalizing $\alpha_t(1), \ldots, \alpha_t(N)$ in each iteration, such that the relative values are preserved, requiring a small modification to the protocol.

11.3.5 Experiments

We performed speech recognition experiments on a 3.2 GHz Intel Pentium 4 machine with 2 GB RAM and running 64-bit GNU/Linux. We created an efficient C++ implementation of the Paillier cryptosystem using the variable precision integer arithmetic library provided by OpenSSL. The encryption and decryption functions were used in a software implementation of the protocol of Sect. 11.2.4.

The experiment was conducted on audio samples of length 0.98 s each. The samples consisted of spoken words which were analyzed by ten HMMs each having 5 states. The HMMs were trained offline on ten different words, viz., utterances of "one", "two", "three", and so on up to "ten". Each test speech sample was converted into 98 overlapping vectors (called frames), a 39-dimensional feature vector composed of MFCCs were extracted from each frame. These MFCC vectors serve as Alice's inputs x_i.

Times needed to process audio samples of length 0.98 s using the 10 HMMs were measured. Table 11.1 gives the processing times for different key lengths with Paillier homomorphic encryption. It is evident that stronger keys incur significant cost in terms of processing time. Note that, once Bob receives the encrypted input from Alice, the calculation of $E(\log b_j(x_t))$ for all 10 HMMs may proceed in parallel. This is possible, for instance, in a cloud computing framework, where Bob has access to several server nodes.

To observe the effect of errors due to finite precision, we compared the values of $\alpha_T(N)$ for a given 0.98-second speech sample derived using the privacy preserving speech recognition protocol as well as a conventional non-secure implementation. This comparison of $\alpha_T(N)$ with and without encryption was performed for each of

Table 11.1 Protocol execution times in seconds for different encryption key sizes

Activity	256-bit keys (s)	512-bit keys (s)	1024-bit keys (s)
Alice encrypts input data (Done only once)	205.23	1944.27	11045.2
Bob computes $E(\log b_j(x_t))$ (Time per HMM)	79.47	230.30	460.56
Both compute $E(\alpha_T(j))$ (Time per HMM)	16.28	107.35	784.58

the 10 HMMs corresponding to the spoken words "one" through "ten". The relative error in $\alpha_T(N)$ was measured to be 0.52 %. Since the error between the actual forward probabilities and their finite precision approximations is very small, the secure speech recognizer nearly always gives the same output as an HMM-based recognizer trained on the same data. Actual performance depends on the quality of the training and test data.

11.4 Discussion

We described protocols for practical privacy preserving inference and classification with HMMs and apply it to the problem of isolated word recognition. A natural progression is to extend the protocols to more complete inference, including decoding from large HMMs with privacy. Other extensions include privacy preserving inference from generalizations of HMMs such as dynamic Bayes networks. The protocols proposed in this chapter represent the first step towards developing a suite of privacy preserving technologies for voice processing, including applications ranging from keyword spotting, to biometrics and full speech recognition.

References

Rane S, Sun W (2010) Privacy preserving string comparisons based on Levenshtein distance. In: Proceedings of IEEE international workshop on information forensics and security (WIFS), Seattle, Dec 2010

Smaragdis P, Shashanka M (2007) A framework for secure speech recognition. IEEE Trans Audio Speech Lang Process 15(4):1404–1413

Part V
Conclusion

Chapter 12
Thesis Conclusion

In this thesis, we introduced the problem of privacy-preserving speech processing. We considered the client/server setting, where the client has the speech input data and the server is interested in performing some speech processing task. We developed various frameworks for performing speech processing on obfuscated data. We considered the problem of privacy-preserving speech processing in the context of three applications: speaker verification, speaker identification, and speech recognition.

We reviewed some of the commonly used algorithms used in the speech processing applications mentioned above: Gaussian mixture models (GMMs) and supervectors for speaker verification and identification, and hidden Markov models (HMMs) for speech recognition. We used the cryptographic techniques such as homomorphic encryption and cryptographic hash functions to create privacy-preserving algorithms. The cryptographic tools that are used in order to preserve privacy introduce an additional computational overhead as compared to the non-private algorithm. Our emphasis is on creating feasible privacy-preserving mechanisms and we evaluate feasibility by measuring accuracy and speed of the mechanisms. We summarize our results below.

12.1 Summary of Results

Part II. In Chap. 4, we considered the problem of speaker verification with the application scenarios of biometric authentication and relevant adversarial models. Conventional speaker verification systems require access to the speaker models and test input samples in plaintext. This makes them vulnerable to attack from an adversary who can gain unauthorized access to the system and use that information to impersonate users. We, therefore, created a framework for privacy-preserving speaker verification, where the system stores obfuscated speaker models and also requires the user to submit obfuscated speech input.

M. A. Pathak, *Privacy-Preserving Machine Learning for Speech Processing*,
Springer Theses, DOI: 10.1007/978-1-4614-4639-2_12,
© Springer Science+Business Media New York 2013

In Chap. 5, we developed SMC protocols for GMM-based algorithm for speaker verification using BGN and Paillier cryptosystems. We constructed both interactive and non-interactive variants of the protocol. The system needs to evaluate the user input on two models: speaker model and UBM. The interactive variant of the protocol is sufficient in the case of semi-honest user that provides correct speech input for both executions of the protocol. To protect against a malicious user that can attempt to cheat by providing different inputs, we use the non-interactive variant of the protocol in which the server takes only one speech input from the user and can apply it to both models without requiring further interaction from the user. The SMC protocols perform the same steps as the original algorithm and obtain identical results up to a high degree of precision. We observed that the interactive and non-interactive protocols achieve high accuracy: 3.1 and 3.8 % EER respectively over the YOHO dataset. In terms of the execution time, the interactive protocol requires 5 min, 33 s and the non-interactive protocol requires 4 h, 52 min to evaluate a 1 s speech sample. The non-interactive protocol is slower because it involves homomorphic multiplication of ciphertexts.

In Chap. 6, we developed a different framework for speaker verification by transforming the problem into string comparison. In this framework, we apply a transformation to convert a speech sample to a fingerprint or a fixed-length bit-string by converting them into supervectors, applying multiple LSH functions, and finally applying a cryptographic hash function to obfuscate the LSH keys. The system then performs then performs speaker verification as password matching, by comparing the input hashes to those submitted by the user during enrollment. Compared to the GMM-based protocols, this framework imposes a very small computational overhead. We observed that the protocol only requires 28.34 ms for each input supervector. In terms of the accuracy, we observe that the algorithm achieves 11.86 % EER. The accuracy is lower than that of using UBM-GMM for speaker verification, but they can potentially be improved by engineering better supervector features. In general, the choice between using LSH and GMM represents a trade-off between execution time and accuracy. If the application deployment scenario requires real-time performance, LSH based speaker verification is more applicable, on the other hand, if the scenario requires high accuracy, it is recommended to use GMM-based speaker verification.

Part III. In Chap. 7, we considered the problem of speaker identification with the application scenario of surveillance. Security agencies perform speaker identification to determine if a given speech sample belongs to a suspect. In order to do so using a conventional speaker identification system, the agency would require access to the speech input in plaintext, By using a privacy-preserving speaker identification system, the agency can perform the identification over obfuscated speech input without violating the privacy of all individuals.

In Chap. 8, we developed the privacy-preserving framework for GMM-based for speaker identification using the Paillier cryptosystem. We constructed two variants of the protocol where either the client sends encrypted speech to the server or the server sends encrypted models to the client. Both the variants of the protocol perform identical computation to the original speaker identification algorithm. We observed accuracy of 87.4 % with a 10-speaker classification task over the YOHO dataset. In terms of the execution time, the two variants require 18 min and 1 h, 22 min respectively to evaluate a 1 s speech sample. The first variant of the protocol requires large bandwidth for long speech inputs, while the bandwidth requirement of the second variant is independent of the speech input length. This represents a trade-off between execution time and bandwidth. In Chap. 9, we developed an algorithm for speaker identification using LSH by transforming the problem into string comparison, similar to Chap. 6. We developed an cryptographic hashing algorithm in which two parties can perform oblivious salting to hide their data from each other. As was the case with speaker verification, the LSH-based protocols have smaller computational overhead as compared to the GMM-based protocols. We observed that the protocol requires 2.9 s for each input supervector. In terms of the accuracy, we observe that the algorithm has 81.3 % accuracy, which is comparatively lower than the GMM-based algorithm. Similar to the case of privacy-preserving speaker verification with LSH, we observe that, the LSH and GMM-based speaker identification algorithms represent a trade-off between execution time and accuracy.

Part IV. In Chap. 10, we introduced the problem of privacy-preserving speech recognition with the application scenario of an external speech service provider. Although external speech processing services have recently become widespread, in many cases, users are reluctant to use them due to concerns about privacy and confidentiality of their speech input.

In Chap. 11, we constructed a privacy-preserving protocols for isolated-word recognition using HMM inference. In this framework HMMs can be evaluated over speech data encrypted with Paillier encryption. By creating one HMM for each word in the dictionary, the server can perform isolated-word speech recognition without observing the client's input. Another advantage of this protocol is that the server is required to perform most of the computational operations as compared to the client. In terms of the execution time, the protocol required 5 min to evaluate a 1 s speech sample.

12.2 Discussion

We created various privacy-preserving frameworks for speech processing tasks discussed above. For each privacy-preserving framework, we performed experiments to calculate the precision and accuracy for the task on a standardized dataset. We compare the execution time of our protocols to that of the non-privacy preserving variant of the same application. In some cases, we created two different mechanisms for the same task, e.g., SMC protocols using homomorphic encryption and string comparison using cryptographic hash functions, that individually provide high accuracy and speed respectively. This provides a trade-off between the two feasibility measures.

The above application scenarios establish the usefulness of our frameworks. Privacy-preserving speaker verification can be used in biometric authentication systems that are robust to being compromised, privacy-preserving speaker identification can be used to perform surveillance with privacy, privacy-preserving speech recognition can be used in external speech services without compromising the confidentiality of the speech input.

In this way, we establish that privacy-preserving speech processing is feasible and useful.

Chapter 13
Future Work

In this thesis, we introduced the problem of privacy-preserving speech processing through a few applications: speaker verification, speaker identification, and speech recognition. There are, however, many problems in speech processing where the similar techniques can be adapted to. Additionally, there are other algorithmic improvements that will allow us to create more accurate or more efficient privacy-preserving solutions for these problems. We discuss a few directions for future research below.

13.1 Other Privacy-Preserving Speech Processing Tasks

13.1.1 Privacy Preserving Music Recognition and Keyword Spotting

Keyword spotting involves detecting the presence of a keyword in an audio sample. A keyword is typically modeled as an HMM which a party can privately train using a audio dataset belonging to another party using a training protocol for speech recognition as described above. Using this model, we can construct an SMC protocol for iterative Viterbi decoding that efficiently finds the sub-sequence an observation sequence having the highest average probability of being generated by a given HMM.

Keyword spotting systems that use word level HMMs are analogous to HMM-based speaker identification systems in their operation, and would require similar approaches to them privately. However, when the system permits the user to specify the words and composes models for them from sub-word units an additional complexity is introduced if it is required that the identity of the keyword be hidden from the system. The approach will require composition of uninformative sub-word graphs and performing the computation in a manner that allows the recipient of the result to obtain scores from only the portions of the graph that represent the desired keyword.

M. A. Pathak, *Privacy-Preserving Machine Learning for Speech Processing*, 117
Springer Theses, DOI: 10.1007/978-1-4614-4639-2_13,
© Springer Science+Business Media New York 2013

13.1.2 Privacy Preserving Graph Search for Continuous Speech Recognition

Another interesting problem is constructing cryptographic protocols for graph search problems. Most problems in speech recognition including model evaluation algorithms like dynamic time warping (DTW), Viterbi decoding, as well as model training algorithms such as Baum-Welch can be formulated as graph search problems; an efficient protocol for privacy preserving graph search can result in practical solutions for many of these problems.

The two specific directions we plan to investigate are:

1. Graph Traversal. We plan to study the general problem where one party has a graph and another party is interested in performing a graph traversal privately. While this is applicable in continuous speech recognition, we also plan to apply it to the general problem of matching finite automata and regular expressions with privacy constraints.
2. Belief Propagation. Given a graph structure, we are interested in evaluating how a potential function such as the probability of having a contagious disease spreads across the nodes in a private manner, i.e., without any node knowing anything about the value of the function at its neighbors. Kearns et al. (2007) present a privacy-preserving message passing solution for undirected trees. We plan to extend it to general graphs.

13.2 Algorithmic Improvements

13.2.1 Ensemble of LSH Functions

In Chaps. 6 and 9, we used LSH as an approximate nearest-neighbor algorithm. In our experiments with LSH, we observed that there is an increase in accuracy by combining LSH for multiple distance metrics. Intuitively, this can be explained by the fact that a single distance metric captures "nearness" in a particular sense. By combining many diverse distance metrics, we are able to capture a heterogeneous sense of nearness that any single distance metric is unable to represent. However, this characteristic is not theoretically well understood. It will be interesting to develop a formal framework to analyze this, including obtaining bounds on the classification error of the ensemble.

13.2.2 Using Fully Homomorphic Encryption

A cryptosystem supporting fully homomorphic encryption (FHE) allows us to perform any operation on plaintext by performing operations on corresponding ciphertext. This finds direct application in creating privacy-preserving mechanisms for distributed computational settings such as cloud-computing. The client can simply encrypt its data with FHE and upload it to the server. As the server can operate on the ciphertext to perform any operation on corresponding the plaintext, we can create a non-interactive protocol for any privacy-preserving computation.

The first such cryptosystem was proposed in a breakthrough work by Gentry (2009, 2010a). Although the construction satisfies the necessary properties for FHE, it is found to be computationally inefficient to be used in practice (Gentry 2010b) in terms of encryption time and ciphertext sizes. Developing practical FHE schemes is an active area of research (Lauter et al. 2011).

If computationally practical FHE schemes are created, we can use them in place of the Paillier and BGN cryptosystems to create non-interactive protocols for speaker verification, speaker identification, and speech recognition. Apart from computing the encrypted probability scores from the encrypted speech input, we can potentially create a non-interactive protocol for private maximum computation.

References

Gentry C (2009) Fully homomorphic encryption using ideal lattices. In: ACM symposium on theory of computing, pp 169–178

Gentry C (2010a) Toward basing fully homomorphic encryption on worst-case hardness. In: CRYPTO, pp 116–137

Gentry C (2010b) Computing arbitrary functions of encrypted data. Commun ACM 53(3):97–105

Kearns M, Tan J, Jennifer W (2007) Privacy-preserving belief propagation and sampling. In: Neural information processing systems

Lauter K, Naehrig M, Vaikuntanathan V (2011) Can homomorphic encryption be practical? In: ACM cloud computing security workshop

Appendix
Differentially Private Gaussian Mixture Models

A.1 Introduction

In recent years, vast amounts of personal data is being aggregated in the form of speech, medical, financial records, social networks, and government census data. As these often contain sensitive information, a database curator interested in releasing a function such as a statistic evaluated over the data is faced with the prospect that it may lead to a breach of privacy of the individuals who contributed to the database. It is therefore important to develop techniques for retrieving desired information from a dataset without revealing any information about individual data instances. *Differential privacy* (Dwork 2006) is a theoretical model proposed to address this issue. A query mechanism evaluated over a dataset is said to satisfy differential privacy if it is likely to produce the same output on a dataset differing by at most one element. This implies that an adversary having complete knowledge of all data instances but one along with *a priori* information about the remaining instance, is not likely to be able to infer any more information about the remaining instance by observing the output of the mechanism.

One of the most common applications for such large data sets such as the ones mentioned above is for training classifiers that can be used to categorize new data. If the training data contains private data instances, an adversary should not be able to learn anything about the individual training dataset instances by analyzing the output of the classifier. Recently, mechanisms for learning differentially private classifiers have been proposed for logistic regression (Chaudhuri and Monteleoni 2008). In this method, the objective function which is minimized by the classification algorithm is modified by adding a linear perturbation term. Compared to the original classifier, there is an additional error introduced by the perturbation term in the differentially private classifier. It is important to have an upper bound on this error as a cost of preserving privacy.

The work mentioned above is largely restricted to binary classification, while multi-class classifiers are more useful in many practical situations. In this chapter, we propose an algorithm for learning multi-class Gaussian mixture model classifiers

M. A. Pathak, *Privacy-Preserving Machine Learning for Speech Processing,*
Springer Theses, DOI: 10.1007/978-1-4614-4639-2,
© Springer Science+Business Media New York 2013

which satisfies differential privacy. Gaussian classifiers that model the distributions of individual classes as being generated from Gaussian distribution or a mixture of Gaussian distributions (McLachlan and Peel 2000) are commonly used as multiclass classifiers. We use a large margin discriminative algorithm for training the classifier introduced by Sha and Saul (2006). To ensure that the learned multi-class classifier preserves differential privacy, we modify the objective function by introducing a perturbed regularization term.

A.2 Large Margin Gaussian Classifiers

We investigate the large margin multi-class classification algorithm introduced by Sha and Saul (2006). The training dataset $(\mathbf{x}, \mathbf{y})$ contains n d-dimensional iid training data instances $\mathbf{x}_i \in \mathbb{R}^d$ each with labels $y_i \in \{1, \dots, C\}$.

A.2.1 Modeling Single Gaussian per Class

We first consider the setting where each class is modeled as a single Gaussian ellipsoid. Each class ellipsoid is parameterized by the centroid $\boldsymbol{\mu}_c \in \mathbb{R}^d$, the inverse covariance matrix $\Psi_c \in \mathbb{R}^{d \times d}$, and a scalar offset $\theta_c \geq 0$. The decision rule is to assign an instance $\mathbf{x}_i$ to the class having smallest Mahalanobis distance (Mahalanobis 1936) with the scalar offset from $\mathbf{x}_i$ to the centroid of that class.

$$y_i = \operatorname*{argmin}_c (\mathbf{x}_i - \boldsymbol{\mu}_c)^T \Psi_c (\mathbf{x}_i - \boldsymbol{\mu}_c) + \theta_c. \tag{A.1}$$

To simplify the notation, we expand $(\mathbf{x}_i - \boldsymbol{\mu}_c)^T \Psi_c (\mathbf{x}_i - \boldsymbol{\mu}_c)$ and collect the parameters for each class as the following $(d + 1) \times (d + 1)$ *positive semidefinite* matrix

$$\Phi_c = \begin{bmatrix} \Psi_c & -\Psi_c \boldsymbol{\mu}_c \\ -\boldsymbol{\mu}_c^T \Psi_c & \boldsymbol{\mu}_c^T \Psi_c \boldsymbol{\mu}_c + \theta_c \end{bmatrix} \tag{A.2}$$

and also append a unit element to each d-dimensional vector $\mathbf{x}_i$. The decision rule for a data instance $\mathbf{x}_i$ simplifies to

$$y_i = \operatorname*{argmin}_c \mathbf{x}_i^T \Phi_c \mathbf{x}_i. \tag{A.3}$$

The discriminative training procedure involves estimating a set of positive semidefinite matrices $\{\Phi_1, \dots, \Phi_C\}$ from the training data $\{(\mathbf{x}_1, y_1), \dots, (\mathbf{x}_n, y_n)\}$ which optimize the performance on the decision rule mentioned above. We apply the large margin intuition about the classifier maximizing the distance of training data instances from the decision boundaries having a lower error. This leads to the classification algorithm being robust to outliers with provably strong generalization guarantees. Formally, we require that for each training data instance $\mathbf{x}_i$ with label y_i, the distance from $\mathbf{x}_i$ to the centroid of class y_i is at least less than its distance from centroids of all other classes by one.

$$\forall c \neq y_i : \mathbf{x}_i^T \Phi_c \mathbf{x}_i \geq 1 + \mathbf{x}_i^T \Phi_{y_i} \mathbf{x}_i. \tag{A.4}$$

Analogous to support vector machines, the training algorithm is an optimization problem minimizing the *hinge loss* denoted by $[f]_+ = \max(0, f)$, with a linear penalty for incorrect classification. We use the sum of traces of inverse covariance matrices for each classes as a *regularization* term. The regularization requires that if we can learn a classifier which labels every training data instance correctly, we choose the one with the lowest inverse covariance or highest covariance for each class ellipsoid as this prevents the classifier from over-fitting. The parameter γ controls the trade off between the loss function and the regularization.

$$J(\Phi, \mathbf{x}, \mathbf{y}) = \frac{1}{n} \sum_{i=1}^{n} \sum_{c \neq y_i} \left[1 + \mathbf{x}_i^T (\Phi_{y_i} - \Phi_c) \mathbf{x}_i \right]_+$$
$$+ \gamma \sum_c \mathrm{trace}(\Psi_c). \tag{A.5}$$

The inverse covariance matrix Ψ_c is contained in the upper left size $d \times d$ block of the matrix Φ_c. We replace it with $I_\Phi \Phi_c I_\Phi$, where I_Φ is the truncated size $(d+1) \times (d+1)$ identity matrix with the last diagonal element $I_{\Phi_{d+1,d+1}}$ set to zero. The optimization problem becomes

$$J(\Phi, \mathbf{x}, \mathbf{y}) = \frac{1}{n} \sum_{i=1}^{n} \sum_{c \neq y_i} \left[1 + \mathbf{x}_i^T (\Phi_{y_i} - \Phi_c) \mathbf{x}_i \right]_+$$
$$+ \gamma \sum_c \mathrm{trace}(I_\Phi \Phi_c I_\Phi). \tag{A.6}$$

The objective function is convex function of positive semidefinite matrices Φ_c. The optimization can be formulated as a semidefinite programming problem (Vandenberghe and Boyd 1996) and be solved efficiently using interior point methods.

A.2.2 Generalizing to Mixtures of Gaussians

We extend the above classification framework to modeling each class as a mixture of K Gaussians ellipsoids. A simple extension is to consider each data instance $\mathbf{x}_i$ as having a mixture component m_i along with the label y_i. The mixture labels are not available *a priori*, these can be generated by training a generative GMM using the data instances in each class and selecting the mixture component with the highest posterior probability. Similar to the criterion in Eq. (A.4), we require that for each training data instance $\mathbf{x}_i$ with label y_i and mixture component m_i, the distance from $\mathbf{x}_i$ to the centroid of the Gaussian ellipsoid for the mixture component m_i of label y_i is at least one greater than the minimum distance from $\mathbf{x}_i$ to the centroid of any mixture component of any other class. If Φ_{y_i,m_i} corresponds to the parameter matrix of the mixture component m_i of the class y_i, and Φ_{cm} corresponds to the parameter matrix of the mixture component m of the class c,

$$\forall c \neq y_i : \min_m \mathbf{x}_i^T \Phi_{cm} \mathbf{x}_i \geq 1 + \mathbf{x}_i^T \Phi_{y_i,m_i} \mathbf{x}_i.$$

In order to maintain the convexity of the objective function, we use the property $\min_m a_m \geq -\log \sum_m e^{-a_m}$ to rewrite the above constraint as

$$\forall c \neq y_i : -\log \sum_m e^{-\mathbf{x}_i^T \Phi_{cm} \mathbf{x}_i} \geq 1 + \mathbf{x}_i^T \Phi_{y_i,m_i} \mathbf{x}_i. \tag{A.7}$$

As before, we minimize the hinge loss of misclassification along with the regularization term. The objective function becomes

$$J(\Phi, \mathbf{x}, \mathbf{y}) = \frac{1}{n} \sum_{i=1}^{n} \sum_{c \neq y_i} \left[1 + \mathbf{x}_i^T \Phi_{y_i,m_i} \mathbf{x}_i + \log \sum_m e^{-\mathbf{x}_i^T \Phi_{cm} \mathbf{x}_i} \right]_+$$
$$+ \gamma \sum_{cm} \text{trace}(I_\Phi \Phi_{cm} I_\Phi). \tag{A.8}$$

After this modification, the underlying optimization problem remains a convex semi-definite program and is tractable to solve. As compared to the single Gaussian case, however, the space of the problem increases linearly as the product of the number of classes and mixture components CK.

A.2.3 Making the Objective Function Differentiable and Strongly Convex

The hinge loss being non-differentiable is not convenient for our analysis; we replace it with a surrogate loss function called Huber loss l_h (Fig. A.1). For small values of the parameter h, Huber loss has similar characteristics as hinge loss and provides the same accuracy (Chapelle 2007). Let us denote $\mathbf{x}_i^T \Phi_{y_i} \mathbf{x}_i + \log \sum_m e^{-\mathbf{x}_i^T \Phi_c \mathbf{x}_i}$ by $M(x_i, \Phi_c)$ for conciseness. The Huber loss ℓ_h computed over data instances $(\mathbf{x}_i, y_i)$ becomes

Fig. A.1 Huber loss

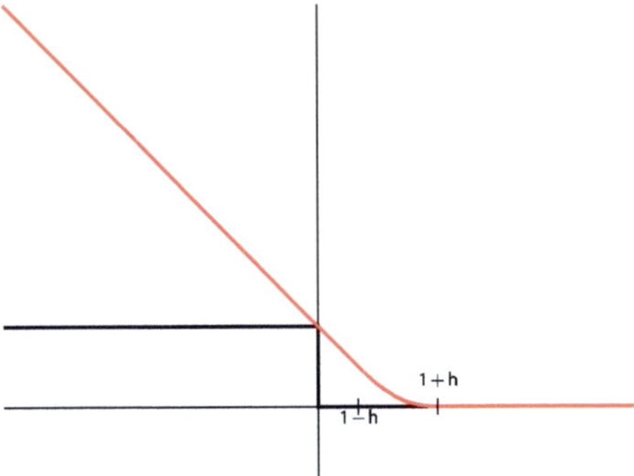

$$
\ell_h(\Phi_c, \mathbf{x}_i, y_i)
$$

$$
= \begin{cases}
0 \\
\quad \text{if } M(x_i, \Phi_c) > h, \\
\frac{1}{4h}\left[h - \mathbf{x}_i^T \Phi_{y_i} \mathbf{x}_i - \log \sum_m e^{-\mathbf{x}_i^T \Phi_c \mathbf{x}_i} \right]^2 \\
\quad \text{if } |M(x_i, \Phi_c)| \le h \\
-\mathbf{x}_i^T \Phi_{y_i} \mathbf{x}_i - \log \sum_m e^{-\mathbf{x}_i^T \Phi_c \mathbf{x}_i} \\
\quad \text{if } M(x_i, \Phi_c) < -h.
\end{cases}
\tag{A.9}
$$

Finally, the regularized Huber loss computed over the training dataset $(\mathbf{x}, \mathbf{y})$ is given by

$$
\begin{aligned}
J(\Phi, \mathbf{x}, \mathbf{y}) = {} & \\
& \frac{1}{n} \sum_{i=1}^{n} \sum_{c \neq y_i} \ell_h \left[1 + \mathbf{x}_i^T \Phi_{y_i} \mathbf{x}_i + \log \sum_m e^{-\mathbf{x}_i^T \Phi_c \mathbf{x}_i} \right] \\
& + \gamma \sum_{cm} \text{trace}(I_\Phi \Phi_{cm} I_\Phi) \\
= {} & \frac{1}{n} \sum_{i=1}^{n} L(\Phi, \mathbf{x}_i, y_i) + N(\Phi) \\
= {} & L(\Phi, \mathbf{x}, \mathbf{y}) + N(\Phi),
\end{aligned}
\tag{A.10}
$$

where $L(\Phi, \mathbf{x}_i, y_i)$ is the contribution of a single data instance to the loss, $L(\Phi, \mathbf{x}, \mathbf{y})$ is the overall loss function and $N(\Phi)$ is the regularization term.

Our theoretical analysis requires that the regularized loss function minimized by the classifier is γ-strongly convex. The regularized loss function $J(\Phi, \mathbf{x}, \mathbf{y})$ is convex as it is the sum of convex loss function $L(\Phi, \mathbf{x}, \mathbf{y})$ and regularization term $N(\Phi)$, but it does not satisfy strong convexity. Towards this, we augment it with an additional ℓ_2 regularization term to get

$$J(\Phi, \mathbf{x}, \mathbf{y}) = \frac{1}{n} \sum_{i=1}^{n} L(\Phi, \mathbf{x}_i, y_i)$$

$$+ \gamma \sum_{cm} \text{trace}(I_\Phi \Phi_{cm} I_\Phi) + \lambda \sum_{cm} \|\Phi_{cm}\|^2$$

$$= L(\Phi, \mathbf{x}, \mathbf{y}) + N(\Phi). \tag{A.11}$$

where $N(\Phi)$ now includes the extended regularization term. As the ℓ_2 regularization term satisfies 1-strong convexity, it can be easily shown that $J(\Phi, \mathbf{x}, \mathbf{y})$ satisfies λ-strong convexity, i.e.,

$$J\left(\frac{\Phi_1 + \Phi_2}{2}, \mathbf{x}, \mathbf{y}\right) = \frac{J(\Phi_1, \mathbf{x}, \mathbf{y}) + J(\Phi_2, \mathbf{x}, \mathbf{y})}{2}$$

$$- \frac{\lambda}{4} \sum_{cm} \|\Phi_{1,cm} - \Phi_{2,cm}\|^2. \tag{A.12}$$

A.3 Differentially Private Large Margin Gaussian Mixture Models

We modify the large margin Gaussian mixture model formulation to satisfy differential privacy by introducing a perturbation term in the objective function. As this classification method ultimately consists of minimizing a convex loss function, the large margin characteristics of the classifier by itself do not interfere with differential privacy.

We generate the size $(d + 1) \times (d + 1)$ perturbation matrix b with density

$$P(b) \propto \exp\left(-\varepsilon \|b\|\right), \tag{A.13}$$

where $\|\cdot\|$ is the Frobenius norm (element-wise ℓ_2 norm) and ε is the privacy parameter. One method of generating such a b matrix is to sample $\|b\|$ from $\Gamma\left((d + 1)^2, \frac{1}{\varepsilon}\right)$ and the direction of b from the uniform distribution.

Our proposed learning algorithm minimizes the following objective function $J_p(\Phi, \mathbf{x}, \mathbf{y})$, where the subscript p denotes privacy.

$$J_p(\Phi, \mathbf{x}, \mathbf{y}) = J(\Phi, \mathbf{x}, \mathbf{y}) + \sum_{c} \sum_{ij} b_{ij} \Phi_{cij}. \tag{A.14}$$

As the dimensionality of the perturbation matrix b is same as that of the classifier parameters Φ_c, the parameter space of Φ does not change after perturbation. In other words, given two datasets $(\mathbf{x}, \mathbf{y})$ and $(\mathbf{x}', \mathbf{y}')$, if Φ^p minimizes $J_p(\Phi, \mathbf{x}, \mathbf{y})$, it is always possible to have Φ^p minimize $J_p(\Phi, \mathbf{x}', \mathbf{y}')$. This is a necessary condition for the classifier Φ^p satisfying differential privacy.

Furthermore, as the perturbation term is convex and positive semidefinite, the perturbed objective function $J_p(\Phi, \mathbf{x}, \mathbf{y})$ has the same properties as the unperturbed objective function $J(\Phi, \mathbf{x}, \mathbf{y})$. Also, the perturbation does not introduce any additional computational cost as compared to the original algorithm.

A.4 Theoretical Analysis

A.4.1 Proof of Differential Privacy

We prove that the classifier minimizing the perturbed optimization function $J_p(\Phi, \mathbf{x}, \mathbf{y})$ satisfies ε-differential privacy in the following theorem. Given a dataset $(\mathbf{x}, \mathbf{y}) = \{(\mathbf{x}_1, y_1), \ldots, (\mathbf{x}_{n-1}, y_{n-1}), (\mathbf{x}_n, y_n)\}$, the probability of learning the classifier Φ^p is close to the probability of learning the same classifier Φ^p given an adjacent dataset $(\mathbf{x}', \mathbf{y}') = \{(\mathbf{x}_1, y_1), \ldots, (\mathbf{x}_{n-1}, y_{n-1})\}$ which *wlog* does not contain the nth instance. As we mentioned in the previous section, it is always possible to find such a classifier Φ^p minimizing both $J_p(\Phi, \mathbf{x}, \mathbf{y})$ and $J_p(\Phi, \mathbf{x}', \mathbf{y}')$ due to the perturbation matrix being in the same space as the optimization parameters.

Our proof requires a strictly convex perturbed objective function resulting in a unique solution Φ^p minimizing it. This in turn requires that the loss function $L(\Phi, \mathbf{x}, y)$ is strictly convex and differentiable, and the regularization term $N(\Phi)$ is convex. These seemingly strong constraints are satisfied by many commonly used classification algorithms such as logistic regression, support vector machines, and our general perturbation technique can be extended to those algorithms. In our proposed algorithm, the Huber loss is by definition a differentiable function and the trace regularization term is convex and differentiable. Additionally, we require that the difference in the gradients of $L(\Phi, \mathbf{x}, y)$ calculated over for two adjacent training datasets is bounded. We prove this property in Lemma 1 given in the appendix.

Theorem 1 *For any two adjacent training datasets $(\mathbf{x}, \mathbf{y})$ and $(\mathbf{x}', \mathbf{y}')$, the classifier Φ^p minimizing the perturbed objective function $J_p(\Phi, \mathbf{x}, \mathbf{y})$ satisfies differential privacy.*

$$\left| \log \frac{P(\Phi^p | \mathbf{x}, \mathbf{y})}{P(\Phi^p | \mathbf{x}', \mathbf{y}')} \right| \le \varepsilon',$$

where $\varepsilon' = \varepsilon + k$ for a constant factor k.

Proof As $J(\Phi, \mathbf{x}, \mathbf{y})$ is strongly convex and differentiable, there is a unique solution Φ^* that minimizes it. As the perturbation term $\sum_c \sum_{ij} b_{ij} \Phi_{cij}$ is also convex and differentiable, the perturbed objective function $J_p(\Phi, \mathbf{x}, \mathbf{y})$ also has a unique solution Φ^p that minimizes it. Differentiating $J_p(\Phi, \mathbf{x}, \mathbf{y})$ wrt Φ_{cm}, we have

$$\frac{\partial}{\partial \Phi_{cm}} J_p(\Phi, \mathbf{x}, \mathbf{y}) = \frac{\partial}{\partial \Phi_{cm}} L(\Phi, \mathbf{x}, \mathbf{y}) + \gamma I_\Phi$$
$$+ 2\lambda \Phi_{cm} + b. \tag{A.15}$$

Substituting the optimal Φ_{cm}^p in the derivative gives us

$$\gamma I_\Phi + b + 2\lambda \Phi_{cm} = -\frac{\partial}{\partial \Phi_{cm}} L(\Phi^P, \mathbf{x}, \mathbf{y}). \tag{A.16}$$

This relation shows that two different values of b cannot result in the same optimal Φ^P. As the perturbed objective function $J_p(\Phi, \mathbf{x}, \mathbf{y})$ is also convex and differentiable, there is a bijective map between the perturbation b and the unique Φ^P minimizing $J_p(\Phi, \mathbf{x}, \mathbf{y})$.

Let b_1 and b_2 be the two perturbations applied when training with the adjacent datasets $(\mathbf{x}, \mathbf{y})$ and $(\mathbf{x}', \mathbf{y}')$ respectively. Assuming that we obtain the same optimal solution Φ^P while minimizing both $J_p(\Phi, \mathbf{x}, \mathbf{y})$ with perturbation b_1 and $J_p(\Phi, \mathbf{x}, \mathbf{y})$ with perturbation b_2,

$$\gamma I_\Phi + 2\lambda \Phi_{cm} + b_1 = -\frac{\partial}{\partial \Phi_{cm}} L(\Phi^P, \mathbf{x}, \mathbf{y}),$$
$$\gamma I_\Phi + 2\lambda \Phi_{cm} + b_2 = -\frac{\partial}{\partial \Phi_{cm}} L(\Phi^P, \mathbf{x}', \mathbf{y}'),$$
$$b_1 - b_2 = \frac{\partial}{\partial \Phi_{cm}} L(\Phi^P, \mathbf{x}', \mathbf{y}') - \frac{\partial}{\partial \Phi_{cm}} L(\Phi^P, \mathbf{x}, \mathbf{y}). \tag{A.17}$$

We take the Frobenius norm of both sides and apply the bound on the RHS as given by Lemma 1. Assuming that $n > 1$, in order to ensure that $(\mathbf{x}', \mathbf{y}')$ is not an empty set,

$$\|b_1 - b_2\| = \left\| \frac{\partial}{\partial \Phi_{cm}} L(\Phi^P, \mathbf{x}', \mathbf{y}') - \frac{\partial}{\partial \Phi_{cm}} L(\Phi^P, \mathbf{x}, \mathbf{y}) \right\|$$

$$= \left\| \frac{1}{n-1} \sum_{i=1}^{n-1} \frac{\partial}{\partial \Phi_{cm}} L(\Phi^P, \mathbf{x}_i, y_i) \right.$$

$$\left. - \frac{1}{n} \sum_{i=1}^{n-1} \frac{\partial}{\partial \Phi_{cm}} L(\Phi^P, \mathbf{x}_i, y_i) - \frac{1}{n} \frac{\partial}{\partial \Phi_{cm}} L(\Phi^P, \mathbf{x}_n, y_n) \right\|$$

$$= \frac{1}{n} \left\| \frac{1}{n-1} \sum_{i=1}^{n-1} \frac{\partial}{\partial \Phi_{cm}} L(\Phi^P, \mathbf{x}_i, y_i) - \frac{\partial}{\partial \Phi_{cm}} L(\Phi^P, \mathbf{x}_n, y_n) \right\|$$

$$\leq \frac{2}{n} \leq 1.$$

Using this property, we can calculate the ratio of densities of drawing the perturbation matrices b_1 and b_2 as

$$\frac{P(b = b_1)}{P(b = b_2)} = \frac{\frac{1}{\text{surf}(\|b_1\|)} \|b_1\|^d \exp\left[-\varepsilon\|b_1\|\right]}{\frac{1}{\text{surf}(\|b_2\|)} \|b_2\|^d \exp\left[-\varepsilon\|b_2\|\right]},$$

where $\text{surf}(\|b\|)$ is the surface area of the $(d+1)$-dimensional hypersphere with radius $\|b\|$. As $\text{surf}(\|b\|) = \text{surf}(1)\|b\|^d$, where $\text{surf}(1)$ is the area of the unit $(d+1)$-dimensional hypersphere, the ratio of the densities becomes

$$\frac{P(b = b_1)}{P(b = b_2)} = \exp\left[\varepsilon(\|b_2\| - \|b_1\|)\right]$$

$$\leq \exp\left[\varepsilon\|b_2 - b_1\|\right] \leq \exp(\varepsilon). \tag{A.18}$$

The ratio of the densities of learning Φ^p using the adjacent datasets $(\mathbf{x}, \mathbf{y})$ and $(\mathbf{x}', \mathbf{y}')$ is given by

$$\frac{P(\Phi^p|\mathbf{x}, \mathbf{y})}{P(\Phi^p|\mathbf{x}', \mathbf{y}')} = \frac{P(b = b_1)}{P(b = b_2)} \frac{|\det(\mathscr{J}(\Phi^p \to b_1|\mathbf{x}, \mathbf{y}))|^{-1}}{|\det(\mathscr{J}(\Phi^p \to b_2|\mathbf{x}', \mathbf{y}'))|^{-1}}, \tag{A.19}$$

where $\mathscr{J}(\Phi^p \to b_1|\mathbf{x}, \mathbf{y})$ and $\mathscr{J}(\Phi^p \to b_2|\mathbf{x}', \mathbf{y}')$ are the Jacobian matrices of the bijective mappings from Φ^p to b_1 and b_2 respectively. In Lemma 3, we show that the ratio of the Jacobian determinants is upper bounded by $\exp(k) = 1 + \frac{1}{n\lambda}$, which is constant in terms of the classifier Φ^p and the dataset $(\mathbf{x}, \mathbf{y})$. The proof of Lemma 3 is similar to Theorem 9 of Chaudhuri et al. (2011).

By substituting this result into Eq. (A.19), the ratio of the densities of learning Φ^p using the adjacent datasets becomes

$$\frac{P(\Phi^p|\mathbf{x}, \mathbf{y})}{P(\Phi^p|\mathbf{x}', \mathbf{y}')} \leq \exp(\varepsilon + k) = \exp(\varepsilon'). \tag{A.20}$$

Similarly, we can show that the probability ratio is lower bounded by $\exp(-\varepsilon')$, which together with Eq. (A.20) satisfies the definition of differential privacy.

A.4.2 Analysis of Excess Error

In this section we bound the error on the differentially private classifier as compared to the original non-private classifier. We treat the case of a single Gaussian per class for simplicity, however this analysis can be naturally extended to the case of a mixture of Gaussians per class. In the remainder of this section, we denote the terms $J(\Phi, \mathbf{x}, \mathbf{y})$ and $L(\Phi, \mathbf{x}, \mathbf{y})$ by $J(\Phi)$ and $L(\Phi)$, respectively for conciseness. The objective function $J(\Phi)$ contains the loss function $L(\Phi)$ computed over the training data $(\mathbf{x}, \mathbf{y})$ and the regularization term $N(\Phi)$—this is known as the regularized *empirical risk* of the classifier Φ. In the following theorem, we establish a bound on the regularized empirical excess risk of the differentially private classifier minimizing the perturbed objective function $J_p(\Phi)$ over the classifier minimizing the unperturbed objective function $J(\Phi)$.

Theorem 2 *With probability at least $1 - \delta$, the regularized empirical excess risk of the classifier Φ^p minimizing the perturbed objective function $J_p(\Phi)$ over the classifier Φ^* minimizing the unperturbed objective function $J(\Phi)$ is bounded as*

$$J(\Phi^p) \leq J(\Phi^*) + \frac{8(d+1)^4 C}{\varepsilon^2 \lambda} \log^2\left(\frac{d}{\delta}\right).$$

Proof We use the definition of $J_p(\Phi) = J(\Phi) + \sum_c \sum_{ij} b_{ij} \Phi_{cij}$ and the optimality of Φ^p, i.e., $J_p(\Phi^p) \leq J_p(\Phi^*)$.

$$J(\Phi^p) + \sum_c \sum_{ij} b_{ij} \Phi^p_{cij} \leq J(\Phi^*) + \sum_c \sum_{ij} b_{ij} \Phi^*_{cij},$$

$$J(\Phi^p) \leq J(\Phi^*) + \sum_c \sum_{ij} b_{ij} (\Phi^*_{cij} - \Phi^p_{cij}). \tag{A.21}$$

Using the strong convexity of $J(\Phi)$ and the optimality of $J(\Phi^*)$, we have

$$J(\Phi^*) \leq J\left(\frac{\Phi^p + \Phi^*}{2}\right)$$

$$\leq \frac{J(\Phi^p) + J(\Phi^*)}{2} - \frac{\lambda}{8} \sum_c \|\Phi^*_c - \Phi^p_c\|^2,$$

$$J(\Phi^p) - J(\Phi^*) \geq \frac{\lambda}{4} \sum_c \|\Phi^*_c - \Phi^p_c\|^2. \tag{A.22}$$

Similarly, using the strong convexity of $J_p(\Phi)$ and the optimality of $J_p(\Phi^p)$,

$$J_p(\Phi^p) \leq J_p\left(\frac{\Phi^p + \Phi^*}{2}\right)$$

$$\leq \frac{J_p(\Phi^p) + J_p(\Phi^*)}{2} - \frac{\lambda}{8} \sum_c \|\Phi^p_c - \Phi^*_c\|^2,$$

$$J_p(\Phi^*) - J_p(\Phi^p) \geq \frac{\lambda}{4} \sum_c \|\Phi^p_c - \Phi^*_c\|^2.$$

Substituting $J_p(\Phi) = J(\Phi) + \sum_c \sum_{ij} b_{ij} \Phi_{cij}$,

$$J(\Phi^*) + \sum_c \sum_{ij} b_{ij} \Phi^*_{cij} - J(\Phi^p) - \sum_c \sum_{ij} b_{ij} \Phi^p_{cij}$$

$$\geq \frac{\lambda}{4} \sum_c \|\Phi^*_c - \Phi^p_c\|^2,$$

$$\sum_c \sum_{ij} b_{ij}(\Phi^*_{cij} - \Phi^p_{cij}) - (J(\Phi^p) - J(\Phi^*))$$

$$\geq \frac{\lambda}{4} \sum_c \|\Phi^*_c - \Phi^p_c\|^2.$$

Substituting the lower bound on $J(\Phi^p) - J(\Phi^*)$ given by Eq. (A.22),

$$\sum_c \sum_{ij} b_{ij}(\Phi^*_{cij} - \Phi^p_{cij}) \geq \frac{\lambda}{2} \sum_c \|\Phi^*_c - \Phi^p_c\|^2,$$

$$\left[\sum_c \sum_{ij} b_{ij}(\Phi^*_{cij} - \Phi^p_{cij}) \right]^2 \geq \frac{\lambda^2}{4} \left[\sum_c \|\Phi^*_c - \Phi^p_c\|^2 \right]^2. \tag{A.23}$$

Using the Cauchy-Schwarz inequality, we have,

$$\left[\sum_c \sum_{ij} b_{ij}(\Phi^*_{cij} - \Phi^p_{cij}) \right]^2 \leq C\|b\|^2 \sum_c \|\Phi^*_c - \Phi^p_c\|^2. \tag{A.24}$$

Combining this with Eq. (A.23) gives us

$$C\|b\|^2 \sum_c \|\Phi^*_c - \Phi^p_c\|^2 \geq \frac{\lambda^2}{4} \left[\sum_c \|\Phi^*_c - \Phi^p_c\|^2 \right]^2,$$

$$\sum_c \|\Phi^*_c - \Phi^p_c\|^2 \leq \frac{4C}{\lambda^2} \|b\|^2. \tag{A.25}$$

Combining this with Eq. (A.24) gives us

$$\sum_c \sum_{ij} b_{ij}(\Phi^*_{cij} - \Phi^p_{cij}) \leq \frac{2C}{\lambda} \|b\|^2.$$

We bound $\|b\|^2$ with probability at least $1 - \delta$ as given by Lemma 6.

$$\sum_c \sum_{ij} b_{ij}(\Phi^*_{cij} - \Phi^p_{cij}) \leq \frac{8(d+1)^4 C}{\varepsilon^2 \lambda} \log^2\left(\frac{d}{\delta}\right). \tag{A.26}$$

Substituting this in Eq. (A.21) proves the theorem.

The upper bound on the regularized empirical risk is in $O(\frac{C}{\varepsilon^2})$. The bound increases for smaller values of ε which implies tighter privacy and therefore suggests a trade off between privacy and utility.

The regularized empirical risk of a classifier is calculated over a given training dataset. In practice, we are more interested in how the classifier will perform on new test data which is assumed to be generated from the same source as the training data. The expected value of the loss function computed over the data is called the *true risk* $\tilde{J}(\Phi) = \mathbb{E}[J(\Phi)]$ of the classifier Φ. In the following theorem, we establish a bound on the true excess risk of the differentially private classifier minimizing the perturbed objective function and the classifier minimizing the original objective function.

Theorem 3 *With probability at least $1 - \delta$, the true excess risk of the classifier Φ^p minimizing the perturbed objective function $J_p(\Phi)$ over the classifier Φ^* minimizing the unperturbed objective function $J(\Phi)$ is bounded as*

$$\tilde{J}(\Phi^p) \leq \tilde{J}(\Phi^*) + \frac{8(d+1)^4 C}{\varepsilon^2 \lambda} \log^2\left(\frac{d}{\delta}\right)$$
$$+ \frac{16}{\lambda n}\left[32 + \log\left(\frac{1}{\delta}\right)\right].$$

Proof Let Φ^r be the classifier minimizing $\tilde{J}(\Phi)$, i.e., $\tilde{J}(\Phi^r) \leq \tilde{J}(\Phi^*)$. Rearranging the terms, we have

$$\tilde{J}(\Phi^p) = \tilde{J}(\Phi^*) + [\tilde{J}(\Phi^p) - \tilde{J}(\Phi^r)] + [\tilde{J}(\Phi^r) - \tilde{J}(\Phi^*)]$$
$$\leq \tilde{J}(\Phi^*) + [\tilde{J}(\Phi^p) - \tilde{J}(\Phi^r)]. \tag{A.27}$$

Sridharan et al. (2008) present a bound on the true excess risk of any classifier as an expression of the bound on the regularized empirical excess risk for that classifier. With probability at least $1 - \delta$,

$$\tilde{J}(\Phi^p) - \tilde{J}(\Phi^r) \leq 2[J(\Phi^p) - J(\Phi^*)]$$
$$+ \frac{16}{\lambda n}\left[32 + \log\left(\frac{1}{\delta}\right)\right].$$

Substituting the bound from Theorem 2,

$$\tilde{J}(\Phi^p) - \tilde{J}(\Phi^r) \leq \frac{8(d+1)^4 C}{\varepsilon^2 \lambda} \log^2\left(\frac{d}{\delta}\right)$$
$$+ \frac{16}{\lambda n}\left[32 + \log\left(\frac{1}{\delta}\right)\right]. \tag{A.28}$$

Substituting this result into Eq. (A.27) proves the theorem.

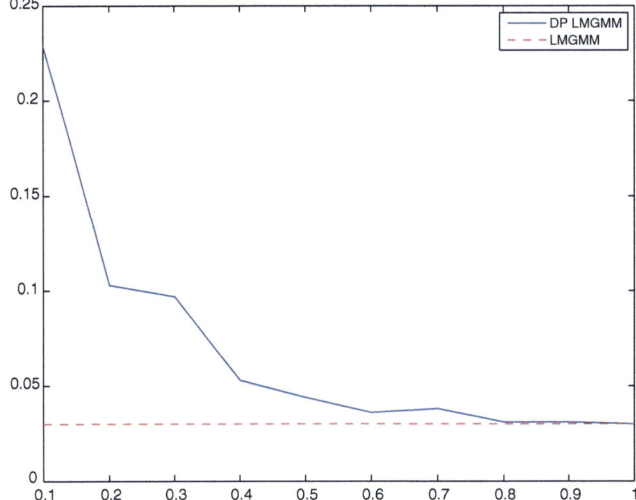

Fig. A.2 Test error versus ε for the UCI breast cancer dataset

Similar to the bound on the regularized empirical excess risk, the bound on the true excess risk is also inversely proportional to ε^2 reflecting the trade-off between privacy and utility. The bound is linear in the number of classes C, which is a consequence of the multi-class classification. The classifier learned using a higher value of the regularization parameter λ will have a higher covariance for each class ellipsoid. This would also make the classifier less sensitive to the perturbation. This intuition is confirmed by the fact that the true excess risk bound is inversely proportional to λ.

A.5 Experiments

We analyze the differentially private large margin Gaussian mixture model classifier to empirically quantify the error introduced by the perturbation. We implemented the classifier using the CVX convex program solver (Grant and Boyd 2010). We report the results on the experiments with the UCI Breast Cancer dataset (Frank and Asuncion 2010) consisting of binary labeled 683 data instances with 10 features. We split the data randomly into 583 instances for the training dataset and 100 instances for the test dataset.

We trained the classifier with the different random samples of the perturbation term b, each sampled with the increasing values of ε, and the regularization parameter $\lambda = 0.31$ which is obtained via cross-validation. The test error results averaged over 10 runs are shown in Fig. A.2.

The dashed line represents the test error of the non-private classifier which remains constant with ε. We observe that for small value of ε implying tighter privacy constraints, we observe a higher error. By increasing ε, we see that the error steadily decreases and converges to the test error of the non-private classifier.

A.6 Conclusion

In this chapter, we present a discriminatively trained Gaussian mixture model based classification algorithm that satisfies differential privacy. Our proposed technique involves adding a perturbation term to the objective function. We prove that the proposed algorithm satisfies differential privacy and establish a bound on the excess risk of the classifier learned by the algorithm which is directly proportional to the number of classes and inversely proportional to the privacy parameter ε reflecting a trade-off between privacy and utility.

A.7 Supplementary Proofs

Lemma 1 *Assuming all the data instances to lie within a unit ℓ_2 ball, the Frobenius norm of the derivative of Huber loss function $L(\Phi, \mathbf{x}_i, y_i)$ computed over the data instance $(\mathbf{x}_i, y_i)$ is bounded.*

$$\left\| \frac{\partial}{\partial \Phi_{cm}} L(\Phi, \mathbf{x}_i, y_i) \right\| \leq 1.$$

Proof The derivative of the Huber loss function for the data instance $\mathbf{x}_i$ with label y_i is

$$\frac{\partial}{\partial \Phi_{cm}} L(\Phi, \mathbf{x}_i, y_i)$$

$$= \begin{cases} 0 \\ \quad \text{if } M(x_i, \Phi_c) > h, \\ \frac{-1}{2h}[h - \mathbf{x}_i^T \Phi_{y_i} \mathbf{x}_i - \log \sum_m e^{-\mathbf{x}_i^T \Phi_c \mathbf{x}_i}] \\ \quad \frac{e^{-\mathbf{x}_i^T \Phi_c \mathbf{x}_i}}{\sum_m e^{-\mathbf{x}_i^T \Phi_c \mathbf{x}_i}} \mathbf{x}_i \mathbf{x}_i^T \\ \quad \text{if } |M(x_i, \Phi_c)| \leq h, \\ \frac{e^{-\mathbf{x}_i^T \Phi_c \mathbf{x}_i}}{\sum_m e^{-\mathbf{x}_i^T \Phi_c \mathbf{x}_i}} \mathbf{x}_i \mathbf{x}_i^T \\ \quad \text{if } M(x_i, \Phi_c) < -h. \end{cases}$$

The data points lie in a ℓ_2 ball of radius 1, $\forall i : \|\mathbf{x}_i\|_2 \leq 1$. Using linear algebra, it is easy to show that the Frobenius norm of the matrix $\mathbf{x}_i \mathbf{x}_i^T$ is same as the ℓ_2 norm of the vector $\mathbf{x}_i$, $\|\mathbf{x}_i \mathbf{x}_i^T\| = \|\mathbf{x}_i\|_2 \leq 1$.

We have,

$$\left\| \frac{e^{-\mathbf{x}_i^T \Phi_c \mathbf{x}_i}}{\sum_m e^{-\mathbf{x}_i^T \Phi_c \mathbf{x}_i}} \mathbf{x}_i \mathbf{x}_i^T \right\| = \left\| \frac{1}{1 + \sum_{m-1} e^{-\mathbf{x}_i^T \Phi_c \mathbf{x}_i}} \mathbf{x}_i \mathbf{x}_i^T \right\|$$

$$= \left| \frac{1}{1 + \sum_{m-1} e^{-\mathbf{x}_i^T \Phi_c \mathbf{x}_i}} \right| \left\| \mathbf{x}_i \mathbf{x}_i^T \right\| \le 1.$$

The term $[h - \mathbf{x}_i^T \Phi_{y_i} \mathbf{x}_i - \log \sum_m e^{-\mathbf{x}_i^T \Phi_c \mathbf{x}_i}]$ is at most h when $|M(\mathbf{x}_i, \Phi_c)| \le h$, the Frobenius norm of the derivative of the Huber loss function is at most one in all cases, $\left\| \frac{\partial}{\partial \Phi_{cm}} L(\Phi, \mathbf{x}_i, y_i) \right\| \le 1$. Similarly, for a data instance $\mathbf{x}_i'$ with label y_i', we have $\left\| \frac{\partial}{\partial \Phi_{cm}} L(\Phi, \mathbf{x}_i', y_i') \right\| \le 1$.

Lemma 2 *Assuming all the data instances to lie within a unit ℓ_2 ball, the Frobenius norm of the second derivative of Huber loss function $L(\Phi, \mathbf{x}_i, y_i)$ computed over the data instance $(\mathbf{x}_i, y_i)$ is bounded.*

$$\left\| \frac{\partial^2}{\partial \Phi_{cm}^2} L(\Phi, \mathbf{x}_i, y_i) \right\| \le 1.$$

Proof The second derivative of the Huber loss function for the data instance $\mathbf{x}_i$ with label y_i is

$$\frac{\partial^2}{\partial \Phi_{cm}^2} L(\Phi, \mathbf{x}_i, y_i)$$

$$= \begin{cases} 0 & \text{if } M(x_i, \Phi_c) > h, \\[2ex] \begin{aligned} &\frac{-1}{2h} \left[\frac{e^{-\mathbf{x}_i^T \Phi_c \mathbf{x}_i}}{\sum_m e^{-\mathbf{x}_i^T \Phi_c \mathbf{x}_i}} \mathbf{x}_i \mathbf{x}_i^T \right]^2 \\ &+ \frac{1}{2h} [h - \mathbf{x}_i^T \Phi_{y_i} \mathbf{x}_i - \log \sum_m e^{-\mathbf{x}_i^T \Phi_c \mathbf{x}_i}] \\ &\frac{e^{-\mathbf{x}_i^T \Phi_c \mathbf{x}_i}}{\left[\sum_m e^{-\mathbf{x}_i^T \Phi_c \mathbf{x}_i} \right]^2} \mathbf{x}_i \mathbf{x}_i^T \end{aligned} & \text{if } |M(x_i, \Phi_c)| \le h, \\[2ex] \frac{e^{-\mathbf{x}_i^T \Phi_c \mathbf{x}_i}}{\left[\sum_m e^{-\mathbf{x}_i^T \Phi_c \mathbf{x}_i} \right]^2} \mathbf{x}_i \mathbf{x}_i^T & \text{if } M(x_i, \Phi_c) < -h. \end{cases}$$

Similar to the arguments we made in Lemma 1, the term $\frac{1}{2h}[h - \mathbf{x}_i^T \Phi_{y_i} \mathbf{x}_i - \log \sum_m e^{-\mathbf{x}_i^T \Phi_c \mathbf{x}_i}]$ is upper bounded by one when $|M(x_i, \Phi_c)| \le h$. As the term $\frac{e^{-\mathbf{x}_i^T \Phi_c \mathbf{x}_i}}{\sum_m e^{-\mathbf{x}_i^T \Phi_c \mathbf{x}_i}} \mathbf{x}_i \mathbf{x}_i^T$ is less than one, its square is also bounded by one.

We apply the Cauchy-Schwarz inequality to the term

$$\frac{e^{-\mathbf{x}_i^T \Phi_c \mathbf{x}_i}}{\left[\sum_m e^{-\mathbf{x}_i^T \Phi_c \mathbf{x}_i}\right]^2} \mathbf{x}_i \mathbf{x}_i^T$$

$$= \frac{1}{1 + \sum_{m-1} e^{-\mathbf{x}_i^T \Phi_c \mathbf{x}_i}} \frac{1}{\sum_m e^{-\mathbf{x}_i^T \Phi_c \mathbf{x}_i}} \mathbf{x}_i \mathbf{x}_i^T$$

$$\leq \left| \frac{1}{1 + \sum_{m-1} e^{-\mathbf{x}_i^T \Phi_c \mathbf{x}_i}} \right| \left| \frac{1}{\sum_m e^{-\mathbf{x}_i^T \Phi_c \mathbf{x}_i}} \right| \|\mathbf{x}_i \mathbf{x}_i^T\| \leq 1,$$

as each of these terms are less than one.

Therefore, the norm of the second derivative is upper bounded by one.

Lemma 3 *The ratio of the determinants of the Jacobian of the mapping $\Phi^P \to b$ computed over the adjacent datasets $(\mathbf{x}, \mathbf{y})$ and $(\mathbf{x}', \mathbf{y}')$ is bounded as*

$$\frac{|\mathscr{J}(\Phi^P \to b | \mathbf{x}', \mathbf{y}')|}{|\mathscr{J}(\Phi^P \to b | \mathbf{x}, \mathbf{y})|} \leq 1 + \frac{1}{n\lambda}.$$

Proof From Eq. (A.16), we have

$$b = -\frac{\partial}{\partial \Phi_{cm}} L(\Phi^P, \mathbf{x}, \mathbf{y}) - \lambda I_\Phi - 2\lambda \Phi_{cm}.$$

We differentiate this wrt Φ_{cm} to get the Jacobian

$$\mathscr{J}(\Phi^P \to b | \mathbf{x}, \mathbf{y}) = \frac{\partial b}{\partial \Phi_{cm}} = -\frac{\partial^2}{\partial \Phi_{cm}^2} L(\Phi^P, \mathbf{x}, \mathbf{y}) - 2\lambda.$$

We represent this by the matrix $-A$. We can similarly write

$$\mathscr{J}(\Phi^P \to b | \mathbf{x}', \mathbf{y}') = -\frac{\partial^2}{\partial \Phi_{cm}^2} L(\Phi^P, \mathbf{x}', \mathbf{y}') - 2\lambda.$$

Noting that the adjacent dataset $(\mathbf{x}', \mathbf{y}')$ is obtained by deleting $(\mathbf{x}_n, y_n)$ from $(\mathbf{x}, \mathbf{y})$, the second derivative of the loss function when evaluated over $(\mathbf{x}', \mathbf{y}')$ can be written as

$$\frac{\partial^2}{\partial \Phi_{cm}^2} L(\Phi^P, \mathbf{x}', \mathbf{y}') = \frac{n}{n-1} \frac{\partial^2}{\partial \Phi_{cm}^2} L(\Phi^P, \mathbf{x}, \mathbf{y})$$

$$- \frac{1}{n-1} \frac{\partial^2}{\partial \Phi_{cm}^2} L(\Phi^P, \mathbf{x}_n, y_n)$$

We can therefore write

$$\mathscr{J}(\Phi^P \to b|\mathbf{x}', \mathbf{y}') = -\frac{n}{n-1}\frac{\partial^2}{\partial \Phi_{cm}^2}L(\Phi^P, \mathbf{x}, \mathbf{y})$$

$$+ \frac{1}{n-1}\frac{\partial^2}{\partial \Phi_{cm}^2}L(\Phi^P, \mathbf{x}_n, y_n) - 2\lambda$$

$$= \frac{1}{n-1}(nA + E)$$

where

$$E = \frac{\partial^2}{\partial \Phi_{cm}^2}L(\Phi^P, \mathbf{x}_n, y_n) + 2\lambda$$

Due to the nature of the Huber loss function, the first term in the RHS is a matrix is of rank at most 1. From Lemma 2, we therefore get $\lambda_1(E) <= 1 + 2\lambda$, where $\lambda_1(E)$ represents the largest Eigenvalue of E.

The ratio of the determinants of the two Jacobians becomes

$$\frac{\mathscr{J}(\Phi^P \to b|\mathbf{x}', \mathbf{y}')}{\mathscr{J}(\Phi^P \to b|\mathbf{x}, \mathbf{y})} = \frac{|\det(A + E)|}{|\det A|}. \tag{A.29}$$

We simplify this ratio of the determinants by applying Lemma 10 of Chaudhuri et al. (2010), which we restate as Lemma 4 below.

$$\frac{|\det(A + E)|}{|\det A|} = |1 + \lambda_1(A^{-1}E) + \lambda_2(A^{-1}E)$$

$$+ \lambda_1(A^{-1}E)\lambda_2(A^{-1}E)|, \tag{A.30}$$

where $\lambda_1(A^{-1}E)$ and $\lambda_2(A^{-1}E)$ are the first and second eigenvalues of the matrix $A^{-1}E$. As $L(\Phi^P, \mathbf{x}, \mathbf{y})$ is a doubly differentiable and a strongly convex function, any eigenvalue of A is at least $n\lambda$ and, therefore, $|\lambda_1(A^{-1}E)| \leq \frac{|\lambda_1(E)|}{n\lambda}$ and $|\lambda_2(A^{-1}E)| \leq \frac{|\lambda_2(E)|}{n\lambda}$.

We use the upper bound on the second derivative of the loss function given by Lemma 2. We apply the triangle inequality to the trace norm, i.e., the sum of the eigenvectors of E, to obtain

$$|\lambda_1(E)| + |\lambda_2(E)| \leq \left\|\frac{\partial^2}{\partial \Phi_{cm}^2}L(\Phi^P, \mathbf{x}_n, y_n)\right\| \leq 1$$

and,

$$|\lambda_1(E)|\,|\lambda_2(E)| \leq 1.$$

We substitute this into the upper bounds for $|\lambda_1(A^{-1}E)|$ and $|\lambda_2(A^{-1}E)|$ and then into Eq. (A.30) to obtain

$$\frac{|\det(A + E)|}{|\det A|} \leq 1 + \frac{1}{n\lambda} + \frac{1}{n^2\lambda^2} \leq \left(1 + \frac{1}{n\lambda}\right)^2.$$

Substituting this bound in Eq. (A.29) proves the theorem. We note that this bound is independent in $(\mathbf{x}, \mathbf{y})$.

Lemma 4 (Lemma 10 of Chaudhuri et al. (2010)) *If A is full rank and E has rank at most 2,*

$$\frac{\det(A + E)}{\det A} = 1 + \lambda_1(A^{-1}E) + \lambda_2(A^{-1}E) + \lambda_1(A^{-1}E)\lambda_2(A^{-1}E).$$

Lemma 5

$$\frac{1}{dC}\left[\sum_c trace[I_\Phi(\Phi_c - \Phi_c')I_\Phi]\right]^2 \leq \sum_c \left\|\Phi_c - \Phi_c'\right\|^2.$$

Proof Let $\Phi_{c,i,j}$ be the (i, j)th element of the size $(d + 1) \times (d + 1)$ matrix $\Phi_c - \Phi_c'$. By the definition of the Frobenius norm, and using the identity $N \sum_{i=1}^N x_i^2 \geq \left(\sum_{i=1}^N x_i\right)^2$,

$$\sum_c \left\|\Phi_c - \Phi_c'\right\|^2 = \sum_c \sum_{i=1}^{d+1} \sum_{j=1}^{d+1} \Phi_{c,i,j}^2 \geq \sum_c \sum_{i=1}^{d+1} \Phi_{c,i,i}^2$$

$$\geq \sum_c \sum_{i=1}^{d} \Phi_{c,i,i}^2 \geq \frac{1}{dC}\left(\sum_c \sum_{i=1}^{d} \Phi_{c,i,i}\right)^2$$

$$= \frac{1}{dC}\left[\sum_c trace[I_\Phi(\Phi_c - \Phi_c')I_\Phi]\right]^2.$$

Lemma 6

$$P\left[\|b\| \geq \frac{2(d + 1)^2}{\varepsilon}\log\left(\frac{d}{\delta}\right)\right] \leq \delta.$$

Proof Similar to the union bound argument used in Lemma 5 in (Chaudhuri and Monteleoni 2008).

References

Chapelle O (2007) Training a support vector machine in the primal. Neural Comput 19(5):1155–1178

Chaudhuri K, Monteleoni C (2008) Privacy-preserving logistic regression. In: Proceedings of neural information processing systems, pp 289–296

Chaudhuri K, Monteleoni C, Sarwate A (2010) Differentially private empirical risk minimization. arXiv:0912.0071v4 [cs.LG]

Chaudhuri K, Monteleoni C, Sarwate A (2011) Differentially private empirical risk minimization. J Mach Learn Res 12:1069–1109

Dwork C (2006) Differential privacy. In: International colloquium on automata, languages and programming

Frank A, Asuncion A (2010) UCI machine learning repository. http://archive.ics.uci.edu/ml

Grant M, Boyd S (2010) CVX: Matlab software for disciplined convex programming, version 1.21. http://cvxr.com/cvx, 2010

Mahalanobis PC (1936) On the generalised distance in statistics. In: Proceedings of the national institute of sciences of india, vol 2, pp 49–55

McLachlan G, Peel D (2000) Finite mixture models. Wiley series in probability and statistics. Wiley, New York

Sha F, Saul LK (2006) Large margin gaussian mixture modeling for phonetic classification and recognition. In: IEEE international conference on acoustics, speech and signal processing, pp 265–268

Sridharan K, Shalev-Shwartz S, Srebro N (2008) Fast rates for regularized objectives. In: Neural information processing systems, pp 1545–1552

Vandenberghe L, Boyd S (1996) Semidefinite programming. SIAM Rev 38:49–95

Author Biography

Dr. Manas A. Pathak received the BTech degree in computer science from Visvesvaraya National Institute of Technology, Nagpur, India, in 2006, and the MS and PhD degrees from the Language Technologies Institute at Carnegie Mellon University (CMU) in 2009 and 2012 respectively. He is currently working as a research scientist at Adchemy, Inc. His research interests include intersection of data privacy, machine learning, speech processing.

Prof. Bhiksha Raj received the PhD degree from CMU in 2000 and was a research scientist at Mistubishi Electric Research Laboratories from 2001–2008. He is an associate professor and nontenured faculty chair at the Language Technologies Institute at Carnegie Mellon University (CMU), and also holds the position of an associate professor by courtesy in the Electrical and Computer Engineering Department at CMU. His chief research interests include robust automatic speech recognition, machine learning and associated topics.

M. A. Pathak, *Privacy-Preserving Machine Learning for Speech Processing*, Springer Theses, DOI: 10.1007/978-1-4614-4639-2,
© Springer Science+Business Media New York 2013